Wastewater Reuse 1

Wastewater Reuse 1

Characteristics, Uses, Applications, Filtration and Disinfection of Water

Kader Gaid

iSTE

WILEY

First published 2025 in Great Britain and the United States by ISTE Ltd and John Wiley & Sons, Inc.

ISTE Ltd
27-37 St George's Road
London SW19 4EU
UK

www.iste.co.uk

John Wiley & Sons, Inc.
111 River Street
Hoboken, NJ 07030
USA

www.wiley.com

Library of Congress Control Number: 2024948091

British Library Cataloguing-in-Publication Data
A CIP record for this book is available from the British Library
ISBN 978-1-78630-979-2

Contents

Chapter 1. Treated Wastewater Reuse: A New Resource 1

1.1. Observations on the current situation. 2
1.2. Climate change. 6
1.3. Solutions to reduce water stress 7
1.3.1. Direct and indirect reuse. 8
1.4. The various purposes of reuse 12
1.5. Adapting to the local context 13
1.6. Policies and institutions to support reuse 14
1.7. Glossary. 15
1.8. References 16

Chapter 2. Characterization of Treated Wastewater 19

2.1. Main parameters defining the quality of treated wastewater 19
2.1.1. Indicators of organic matter: biological oxygen demand, chemical oxygen demand and total organic carbon 21
2.1.2. Suspended solids 25
2.1.3. Nitrogen 26
2.1.4. Phosphorus. 27
2.1.5. Pesticides and endocrine disruptors 28
2.2. Microbiological aspects 33
2.2.1. Bacteria. 35
2.2.2. Enteric viruses. 37
2.2.3. Enteric protozoa. 38
2.2.4. Helminth eggs 38
2.2.5. Fecal contamination indicators 40
2.3. References 45

Chapter 3. Applications and Uses of Reused Water 49

3.1. The different uses 54
3.1.1. Agriculture. 54
3.1.2. Reuse in industry 67
3.1.3. Groundwater recharge 69
3.1.4. Recreational and environmental areas. 71
3.1.5. Urban non-food uses 72
3.1.6. Drinking water 74
3.2. References 75

Chapter 4. Regulations 85

4.1. Regulations, state of the art and future challenges of municipal wastewater treatment and water reuse 85
4.2. Regulations for treated wastewater reuse in France 87
4.2.1. Application to irrigation 89
4.3. European Directive 95
4.3.1. Minimum requirements applicable to reclaimed water intended for agricultural irrigation. 97
4.4. World Health Organization 100
4.4.1. Objectives of pathogen reduction for the different irrigation types 103
4.4.2. Monitoring/verification 108
4.5. Regulation on treated wastewater reuse in the United States. 108
4.6. Regulation on water reuse in China. 112
4.7. Australia. 113
4.8. Mediterranean countries and the Middle East. 115
4.9. Conclusion 116
4.10. References. 121

Chapter 5. Treated Wastewater Reuse Planning. 127

5.1. Project objectives and limitations. 127
5.1.1. Preliminary investigations. 128
5.1.2. Selection of potential markets 128
5.1.3. Assessment of alternatives 129
5.1.4. Awareness and education 130
5.1.5. Importance of terminology 131
5.1.6. Economic and legal aspects 131
5.2. References 132

Chapter 6. Treatment Technologies 133

6.1. Use of raw wastewater 135
6.2. Bases of the treatment process concept 136
6.3. References 139

Chapter 7. Tertiary Filtration 141

7.1. Gravity filtration without reagents 142
7.1.1. Rapid filters and high-rate filters (HRF) 142
7.1.2. Dual-media filters 144
7.1.3. Pressure filters 146
7.2. Gravity filtration with reagents 147
7.3. Filtration mechanisms 148
7.4. Implementation parameters 150
7.4.1. Media 150
7.4.2. Media height ratio/d_{10} 157
7.4.3. Effective size ratio 158
7.5. Size parameters: filtration rate and media height 159
7.5.1. Filter depth modelling 160
7.6. Operating parameters 162
7.6.1. Pressure drop in a clean filter 162
7.6.2. Retention capacity 165
7.6.3. Filter backwashing conditions 168
7.7. Filtration technologies 178
7.7.1. Conventional single-layer filters 180
7.7.2. High-rate filters 182
7.7.3. Example of tertiary filtration design on single-media filter 187
7.7.4. Dual-media filters 188
7.7.5. Filters under pressure 194
7.7.6. Performances of gravity filters 196
7.8. Mobile material filters 197
7.9. Rotary drum filters 198
7.9.1. Hydrotech filter characteristics 200
7.10. References 202

Chapter 8. Disinfection 207

8.1. Microorganisms present in treated wastewater 207
8.1.1. Bacteria 208
8.1.2. Viruses 210
8.1.3. Parasites 211

8.2. General rules of chemical disinfection 212
8.2.1. Mode of action of chemical disinfectants in treated wastewater 212
8.2.2. Reduction principle of the number of germs – "Log" units 214
8.3. Factors influencing the effectiveness of chemical disinfection. 217
8.3.1. Turbidity 218
8.3.2. Presence of oxidizable matter 218
8.3.3. Injection mode and point 218
8.3.4. Design of the contact tank 219
8.4. Qualities of a good disinfectant 219
8.5. The main techniques of wastewater chemical disinfection 220
8.5.1. Chlorine disinfection 220
8.5.2. Chlorine dioxide disinfection 243
8.5.3. Chloramination 253
8.6. Disinfection with ozone 263
8.6.1. General information on ozone 264
8.6.2. Preparation and implementation of ozone 265
8.6.3. Transfer of ozone in water to be treated. 266
8.6.4. Performance of disinfection with ozone. 268
8.7. Organic peracids: performic acid and peracetic acid. 271
8.7.1. Production organic peracids. 271
8.7.2. Disinfection and oxidation mechanisms of organic peracids 272
8.7.3. Performic acid 274
8.7.4. Peracetic acid 276
8.7.5. Degradation kinetics of peracids 276
8.7.6. Primary, secondary and tertiary wastewater applications 278
8.8. UV disinfection 281
8.8.1. General information on UV radiation 281
8.8.2. Inactivation mechanisms 282
8.8.3. Lethal dose and inactivation kinetics 284
8.8.4. Implementation 286
8.8.5. Sizing parameters 290
8.8.6. Factors influencing UV treatment efficiency. 291
8.8.7. Performance 295
8.8.8. UV disinfection advantages and disadvantages in wastewater 300
8.9. Criteria for selecting a disinfection technique. 302
8.10. References 304

Index 313

1

Treated Wastewater Reuse: A New Resource

Throughout the world, the availability of food and water is vital for the survival of humanity and to ensure an adequate standard of living. Increasing population, climate change and ongoing industrialization are putting pressure on existing water resources. Freshwater resources are not sufficient in all regions of the world. It is estimated that 50% of the world's population will live in water-stressed regions by 2030, underlining the importance of sound management and adequate water treatment. When water resources are limited, irrigation is severely penalized, as agriculture uses 70% of the total amount of water taken.

The water issue is inseparable from sustainable development, in that water must be able to meet the needs of present generations without jeopardizing the ability of future generations to satisfy their own. With the consequences of climate change, pressure on resources will continue to increase under the combined effects of population growth and policies implemented in respect of water-consuming activities. The water deficit will ipso facto lead to serious conflicts between the various users, which will inevitably require difficult arbitrations for public authorities. In many parts of the world, the 21st century is set to see a worsening of water shortages. According to the UNDP (United Nations Development Programme), an increase in global temperature of 3–4°C would cause a drought that would affect hundreds of millions of people. The IPCC (Intergovernmental Panel on Climate Change) confirms that the scale of impacts on water shortages would be dramatic.

For a color version of all figures in this chapter, see www.iste.co.uk/gaid/reuse1.zip.

Thus, water scarcity will affect not only irrigation for human food but also animal production (where water is essential both for drinking and feeding).

In order to improve agricultural production and minimize the depletion of natural water resources, a targeted low-cost water quality is needed. Wastewater reuse is a first solution.

1.1. Observations on the current situation[1]

The current situation is further exacerbated by several constraints: an imbalance between needs and available resources, a considerable increase in the needs for drinking, industrial and agricultural water, a geographical imbalance between needs and resources, pollution of groundwater and surface water and significantly reducing in some countries the volumes of water that can be used. The risk of a break in sustainable development may arise if the actual groundwater withdrawals exceed the limits for renewal of natural resources.

In Spain, April 2023 was one of the hottest and driest months recorded. Between the lack of fodder and soaring food prices, farmers had no choice but to slaughter their cattle.

In 2022, a prolonged drought affected many regions of France, Italy and Spain. The lack of precipitation combined with heat waves has led to dried out soils and a lack of water. The socioecological impacts of this drought were dramatic. The high water and thermal stress have significantly reduced yields of major crops such as corn, soybeans and sunflower, with reductions of about 15% in comparison to the last 5 years' average.

In Italy, about 50% of the population has been affected by water supply restrictions. High levels of salt intrusion from the Po Delta up to 40 km from the maritime coast have been recorded. The sharp drop in water levels in dams has severely impacted hydroelectric power and cooling systems of power plants.

In the south of France, fires associated with extreme drought conditions doubled compared to 2021 and were more than four times higher than during the 2012–2021 period. Sixty-six French departments were at the highest alert level in August, and

1 A few technical and scientific descriptions presented in this book were partly taken from the book *Drinking Water Treatment* (Gaid 2022), as the processes used in both applications were very similar.

93 departments were affected by all three alert levels. Similar impacts were observed in Spain and Portugal for agriculture, energy production and domestic water use.

The 2022 drought had significant societal impacts and the fear of future droughts is a major concern for populations (Breda and Badeau 2008; Baum et al. 2013; Faranda et al. 2023). According to the World Meteorological Organization, drought is one of the most damaging and deadly climate risks. In the coming years, drought-related rainfall deficits will become very problematic.

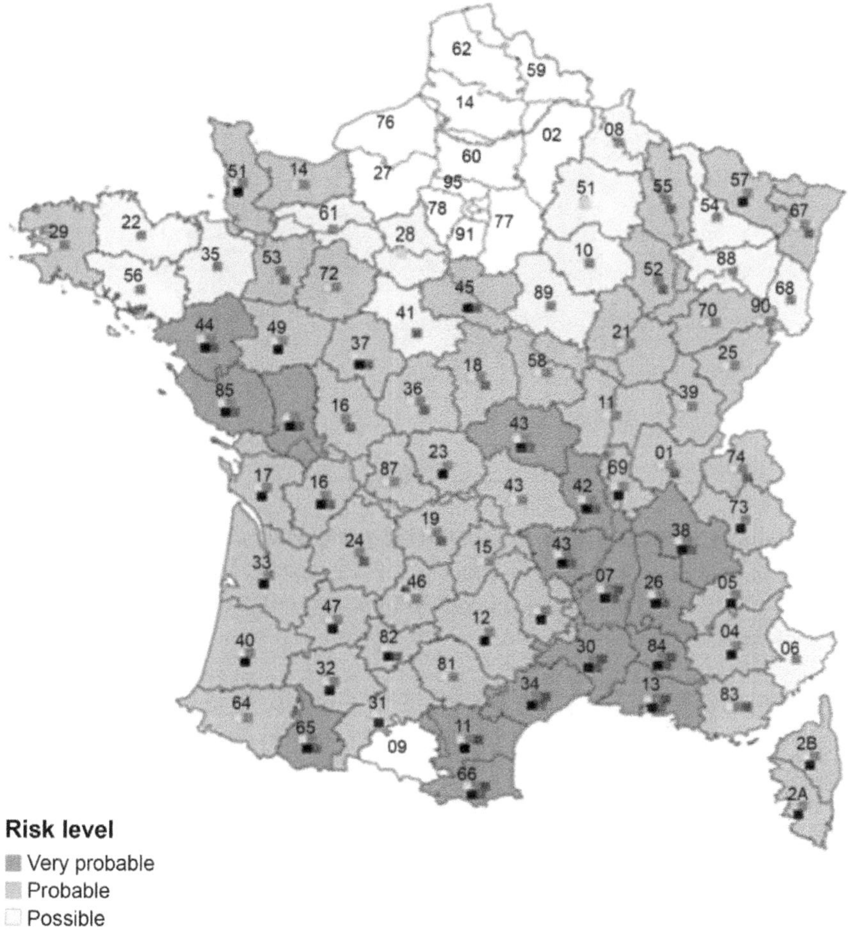

Figure 1.1. *(A) Maximum severity level limiting water use by department in force during the 2021 summer*

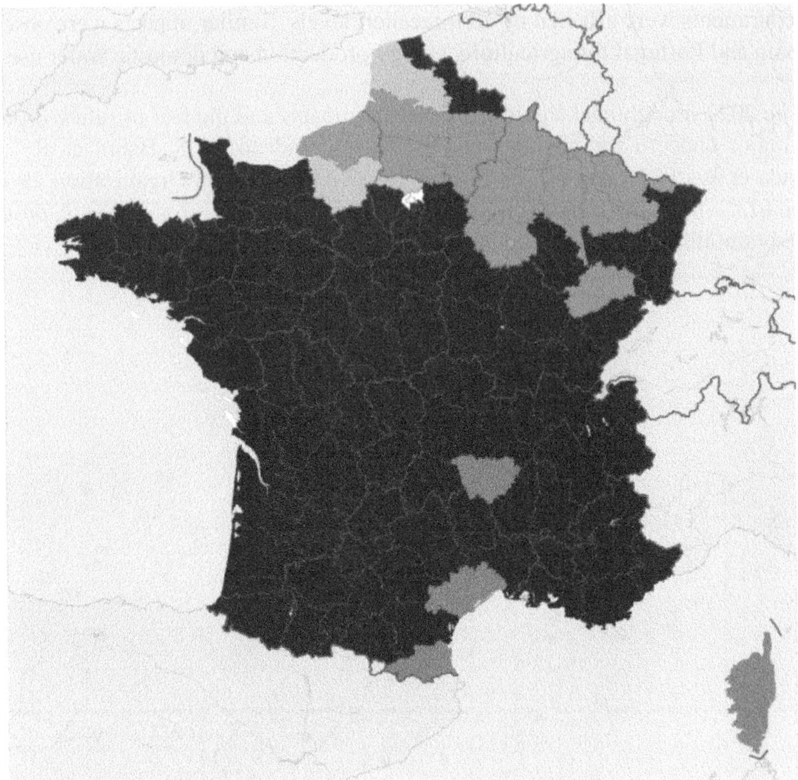

Figure 1.1. *(B) Maximum severity level limiting water use by department in force (continued) as of August 12, 2022 (source: Ministère de la Transition écologique et de la Cohésion des territoires)*

Figure 1.2. *Worsening of soil condition (France)*

Figure 1.3. *Lack of foods drives Spanish cattle farmers to massively kill cattle*

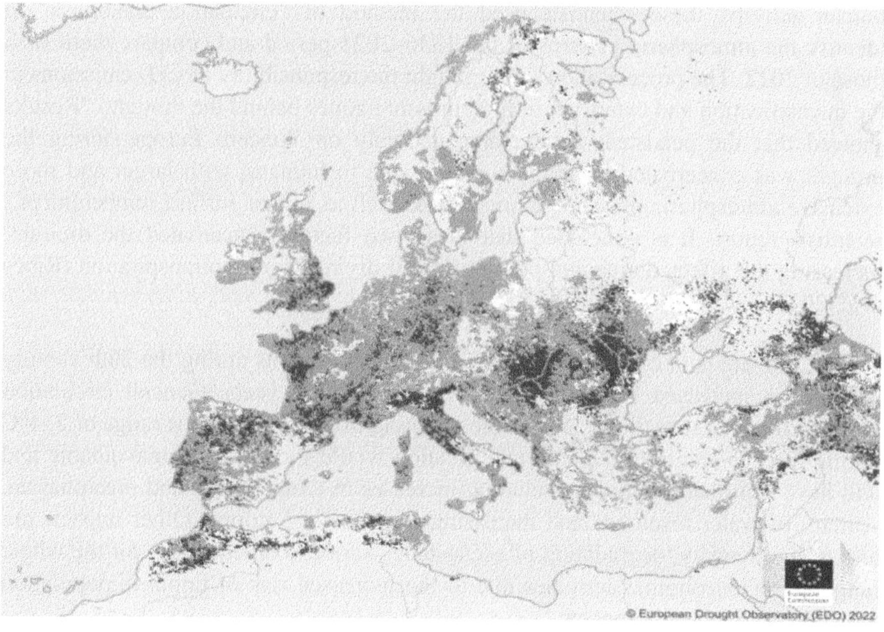

Figure 1.4. *Drought in Europe and North Africa (2022)*

The question of how much water resources will be available in the coming decades will be determined by climate change, even if water quality will depend on other factors such as infrastructure management, perimeter protection and pollution control. The expected and already observed summer drying in many countries has resulted in a number of impacts: tree health, natural resources, agriculture, water resources and hydroelectric power supply. Many regions of the world are concerned with the risk of increased frequency and intensity of heat waves and droughts.

1.2. Climate change

The role of climate change in the prolonged drought that affected Western Europe and the Mediterranean region in 2022 is now well established. Published in the journal *Environmental Research Letters*, a study conducted by a CNRS (National Centre for Scientific Research) team has established links between climate change and the occurrence of drought episodes, such as those encountered by France during the summer of 2022.

To demonstrate the responsibility of "anthropogenic climate change", linked to human activity, these scientists used the method of "circulation analogues" to identify the atmospheric patterns of the 1836–2021 period and compare them with those of 2022. The process allows to highlight the responsibility of CO_2 emissions in the intensification and extension of high pressure zones behind the drought. "Results showed that the persistent anticyclonic anomaly on Western Europe during the drought was exacerbated by climate change due to humans, with larger and more extensive atmospheric pressure anomalies, as well as higher surface temperatures", scientists report. It is concluded that these two factors exacerbated the drought, increasing the affected area and promoting soil drying by evapotranspiration (Ribes et al. 2022).

The climate data collected in the Mediterranean regions during the 20th century indicate an increased warming trend over the last 30 years. General circulation models converge to estimate a probable warming of the region in the range of 2–4°C during the 21st century. Climate modification would therefore be unavoidable and will have significant impacts, including increases in temperatures and precipitation, scarcity of water resources and increasing frequency of storms. Other impacts are loss of biodiversity, degradation of ecosystems, serious consequences for the whole range of socioeconomic activities due to the increased risk of famines, population movements and health impacts.

These changes will affect the retention of surface water and the useful capacity of dams, the renewal of groundwater levels. The latter point has already shown that drought combined with overexploitation of groundwater can lead to mineralization of unsaturated areas of deep aquifers. In coastal regions, the drop in hydrostatic pressure levels has already led to sea water entering the freshwater reserves of coastal aquifers. Climate change is having an impact on public health with potential for outbreaks of water-borne diseases. In North Africa, for example, this is due to insufficient water resources combined with the lack of treatment of some water points.

1.3. Solutions to reduce water stress

Some countries, such as the United States, Singapore and Namibia, have been pioneering for decades the implementation of the concept of treated wastewater reuse with alternatives for reusing drinking and non-drinking water. Wastewater treatment technologies currently provide solutions for adequate supply of reclaimed water. Therefore, the opportunities for water reuse exist, but the challenge lies largely in implementation. This is becoming urgent and necessary in regions where water stress is most pressing. Climate change will worsen the situation and drought periods will multiply. Total water demand will increase, especially for irrigation of field crops. As a result, water from natural resources such as groundwater and surface water will often be over-used.

The apprehension of climate change requires the implementation of strategies and controlled planning to satisfy water supplies. Only then will it be possible to meet the challenge of population growth and ensure safe drinking water for all. Alternative water sources must be produced to meet this challenge, as with drought, groundwater and surface water availability is already often insufficient and will be of great concern in many parts of the world, especially since the reduction of rain periods does not allow to fill dams, hill reservoirs and retention ponds.

Several solutions exist to reduce water stress with non-conventional waters:

– the desalination of non-conventional waters (sea water, oceans, estuaries, etc.);

– the use of fossil (non-renewable) groundwater;

– the reuse of wastewater.

Desalination is a good, very interesting, more expensive solution but primarily ensures the supply of coastal cities.

The use of fossil groundwater is another solution sometimes used, which has the major inconvenience of draining these reservoirs that are not renewable. Ground collapses were also observed when implementing this solution.

Figure 1.5. *Desalination and/or treated wastewater reuse*

Compared to desalination, even if drinking water applications, for example, are more reliable and reassuring, another alternative and sustainable resource is the reuse of treated wastewater. It stands as a new resource that can be directly available without passing through receiving environments. Water reuse is a cost-effective and energy-efficient option to increase water supply and mitigate the impact of climate variability and climate change. For all these reasons, treated wastewater reuse has emerged as the most promising, accessible and easy-to-implement solution. It is a voluntary and planned action, which aims at the production of additional quantities of water for different uses in order to fill water deficits.

1.3.1. *Direct and indirect reuse*

Direct reuse and indirect reuse have to be distinguished.

Direct reuse of treated wastewater is not to discharge wastewater into a river after treatment, but to immediately reinject it in a network for dedicated use (irrigation, watering green spaces or golf courses, cleaning roads, etc.). Direct reuse in France related to only 1% of wastewater, compared to 8% in Italy and 14% in Spain. This practice is highly developed in Israel, where 90% of wastewater is reused, without passing again in streams, most of them intermittent, meaning that these rivers only flow part of the year (BRGM 2010; Cerema 2020; Onema 2022).

Direct drinking water reuse is the supply of reused water that has undergone advanced complementary treatment and is sent directly to a drinking water treatment plant or distribution system, with or without intermediate storage. The potential barriers or disadvantages of this option are mainly related to public perception and acceptance rather than science or engineering.

Direct reuse is banned in France for drinking water and is not authorized for most food industries for health reasons.

Indirect reuse comes to draw water from a stream or its alluvial aquifer, while at a short distance upstream, a plant has discharged wastewater after treatment. The treated water has been diluted in the receiving medium. Indirect reuse is widespread in summer in all streams in France, where most irrigation is done by pumping rivers or alluvial aquifer. In other words, a significant part of the wastewater discharged in summer is already being reused in France or is used to maintain low water flow rates, except when stations discharge near the seafront or into the sea. Indirect reuse is permitted for all uses (BRGM 2010; Cerema 2020; Onema 2022).

This solution is widely practiced in many parts of the world and implies the availability of an environmental buffer zone (such as a lake, reservoir or aquifer). The water taken from this environment is then sent and injected upstream of the drinking water treatment plant.

Water reuse (also known as water recycling or water recovery) involves recovering water from various sources, and then treating and reusing it for beneficial purposes, such as agriculture and irrigation, drinking water supply, groundwater replenishment, industrial processes and environmental restoration. Water reuse can provide alternatives to existing water supplies and be used to improve water safety, durability and resilience.

Water reuse can be defined as planned or unplanned. Unplanned water reuse refers to situations in which a water source is essentially made up of previously used water. A common example of unplanned water reuse occurs when communities source water from streams, such as the Colorado and Mississippi rivers, which receive treated wastewater discharges from upstream communities.

Planned water reuse refers to water supply systems designed to advantageously reuse a recycled water supply source. Often, communities will seek to optimize their overall water use by reusing water as much as possible within the community, before the water is returned to the environment. Examples of planned reuse include agricultural and landscape irrigation, industrial process water, drinking water supply and groundwater supply management.

All water treatment facilities require a high level of reliability to ensure quality that meets standards and minimize, or even eliminate, any public health risk. This principle is particularly important in the case of RDP, where the risks of contaminants leaking after technical failures must be carefully controlled.

Three main water sources are the subject of these applications: water from domestic wastewater treatment plants, urban rainwater and industrial wastewater.

Direct reuse in coastal areas allows an additional valorization of wastewater whose discharge generates pollution (bathing water, risks for shellfish aquaculture, etc.).

The sprinkling of golf courses in France has sometimes reduced water withdrawals from communities. Similarly, in all contexts, direct reuse in town has potential (cleaning of roads and street furniture, green spaces or industry).

The reuse of treated wastewater is also considered when there is a need to protect a receiving environment, whether marine or underground, or when there is a water stress situation, which makes it necessary to conserve water resources.

The development of a territory, combined with urban, peri-urban or rural development projects, may involve recycling treated wastewater for the purpose of watering recreational areas, of agricultural or industrial use.

In most cases, the treated wastewater is discharged into natural watercourses where it undergoes further self-purification. They are then pumped to make drinking water for various industrial or agricultural uses. The receiving environment is known to promote the reduction of biodegradable pollutants in treated waters and, by dilution, reduces the impact of toxic elements. This last effect is highly limited, even unlikely, when the environment is water deficient.

The advantage of reuse is to create a shortcut for the accumulated mechanisms within the environment and to avail quickly of a water resource. In this case, in connection with the quality of the water obtained and the desired use, the availability of this new resource opens up immediate prospects.

Reuse of wastewater for direct consumption can raise public health and possibly religious concerns among consumers. Reuse as a means of freshwater supply is determined by economic and health factors involving water stress.

In North Africa, the Sahel and the Middle East, strong specificities govern water resources and their use, as these countries have mostly deserts that do not allow for renewable resources. Water stress in these countries creates complicated situations for agriculture and industry. Desalination plants have been installed where possible, but do not solve the problem in its entirety. The deficit remains high in desert areas.

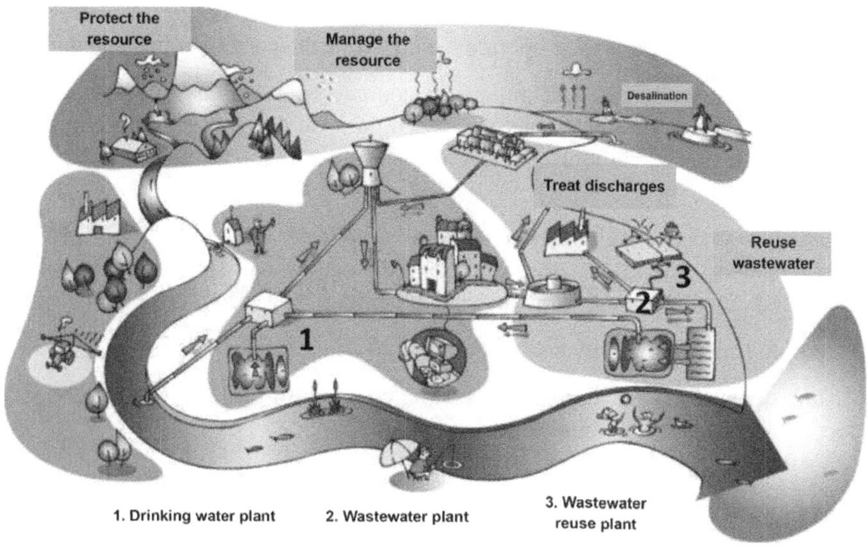

Figure 1.6. *Global water cycle after treated wastewater reuse*

In all the countries of the world where water stress has emerged, government strategies and policies have been focused on resource management and protection for the last 15 years, and also, in numerous cases on treated wastewater reuse.

Thus, in a world where cities are growing, and climate change threatens to exacerbate water scarcity in many parts of the world, treated wastewater reuse should be an integral part of water resource management, water supply and sanitation strategies.

To make wastewater an asset, an integrated approach that encompasses the entire water cycle – available water resources and supply options, water treatment and reuse – must be adopted.

As such, it is often recommended to:

– consider wastewater as a potential resource in the overall water balance;

– adopt an integrated framework to manage water supply, rainwater, wastewater, nonpoint sources, pollution and water reuse;

– integrate wastewater recovery and reuse into sustainable development and climate change as integrated water resource adaptation and management strategies;

– consider different reuse options from the outset in the design of wastewater treatment plants, as well as their operation, and define the corresponding standards;

– ensure that guidelines and policies encourage communities to identify cost-effective wastewater treatment solutions based on local capacity and reuse options, rather than imposing solutions or severely limiting them through overly strict regulations;

– involve all stakeholders from the outset in water reuse plans and provide multistakeholder platforms to facilitate dialogue, participatory technology development, innovation adoption and social learning;

– ensure financial stability and sustainability;

– link waste management to other economic sectors for faster cost recovery, risk reduction and sustainable implementation;

– develop public/private, public/public mixed solutions for investment, service delivery, operation and maintenance;

– take social equity into account when defining cost recovery mechanisms.

Climate change requires us to change our mentality and see wastewater as an alternative resource and an economic and environmental opportunity. Better wastewater management would actually help reduce water deficit, water scarcity and pollution of receiving environments.

1.4. The various purposes of reuse

Reusing a community's wastewater involves collecting, storing and using the wastewater for various purposes.

The various applications of treated wastewater reuse can be summarized as:

– urban environment (watering public parks, green spaces, sports fields, supply of water features and fountains, watering of private gardens, toilet flushing, washing of vehicles, street cleaning, fire circuits, etc.);

– industrial environment (cooling water, process water, etc.);

– agricultural (irrigation of various crops, market gardening, fruit trees, cereals, fodder, pastures, industrial crops and forests) and underground (groundwater recharge) environment;

– production of drinking water (groundwater recharge, direct production of drinking water);

– aquaculture;

– increase in drinking water;

– groundwater recharge;

– restoration of water bodies and wetlands;

– recreation areas.

Water reuse is already a widespread and accepted practice in many countries. Depending on its applications, the population may have negative opinions of it. Protests have been held in the states of a few countries against the use of recycled water for domestic purposes.

However, some countries have a long history of planned reuse and wastewater treatment plants producing water for various uses, including drinking water. In Namibia, South Africa and the United States, the strategies in place cover wastewater treatment – across a range of conventional and non-conventional systems – and national guidelines and regulations for reuse.

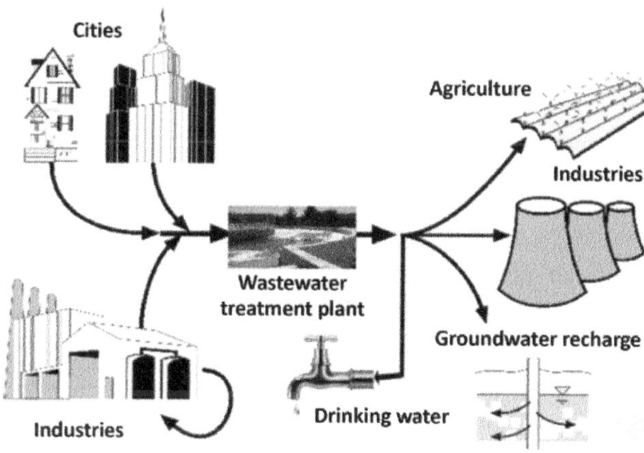

Figure 1.7. *Applications of treated wastewater reuse*

1.5. Adapting to the local context

In addition to treating wastewater as an asset, it is important that changes be made in a number of key areas for effective management of the global water cycle. For example:

– defining the most suitable and viable option for selected applications;

– taking into account local capabilities for operation, maintenance and downstream uses. Decentralized systems can better protect watersheds and water resources, and avoid long-distance wastewater transfers;

– managing water, wastewater, pollution control and water reuse in an integrated way within a comprehensive program;

– preferably, planning for the location of treatment to be within the vicinity of reuse sites;

– defining appropriate and cost-effective processing levels to match each reuse option;

– ensuring that water production meets water quality requirements for the specified use. Considering reuse from the outset in the design of wastewater treatment plants, as well as their operation, will provide the best results;

– changing to an approach where the treatment of raw wastewater and the complementary treatment for water reuse are combined in the same site. This helps to ensure the overall operation of the system, control the water quality and intervene very quickly in case of failure.

1.6. Policies and institutions to support reuse

The issue of treated wastewater reuse involves changes in our daily management of water consumption. For example, reinventing water efficiency, adopting innovative business models and mobilizing the water community.

In order to ensure public acceptance of water reuse, political will must be strengthened through constant information:

– The legislation may therefore have to define the access rights of users and the powers of the right holders. To ensure safe reuse, it is important to make access dependent on farmers' compliance with the reuse guidelines.

– Implementing economic incentives: incentives to reuse treated wastewater are useful where water users can choose between different sources of water.

– Bridging sectoral and administrative divides: effective wastewater management requires cooperation between agencies and sectors, such as health, municipal wastewater treatment, irrigation water distribution, etc.

– To get public understanding and acceptance, regulations on water reuse must include clear texts applying to drinking water, bathing water, irrigation water, drainage, etc.

– Public awareness, education and transparency: public awareness and education programs, including school programs, should be promoted.

– Transparency, information sharing and the involvement of water users in the decision-making process will also ensure greater acceptance of reuse projects. In the case of water reuse in agriculture, farmers must be educated on safe irrigation and post-irrigation practices, and consumers must be informed of the safety of agricultural products irrigated with well-managed recovered water. Quality data on water must be widely available and freely shared with water customers and the general public.

– Stakeholder participation should be encouraged as it is essential to success. Communities must be able to express their needs and suggestions in open multi-stakeholder platforms.

– Long-term strategies for progressive action and new local business models are also needed.

The reuse of wastewater is currently a socioeconomic issue for the development of drinking water and sanitation services. It has the advantage of providing an alternative resource to limit water deficits, better preserve natural resources and mitigate water shortages caused by climate change. Already, some countries, states and major cities (Australia, California, Cyprus, Spain, Florida, Israel, Jordan, Malta, Singapore, etc.) have ambitious targets of meeting 10–30% or even 60% of their water demand through the reuse of treated wastewater (Degrémont 2005).

This book presents wastewater characteristics, health risks, reuse regulations, various uses of reused water and the performance of conventional and membrane sectors: the ability to remove pollutants, the elimination of micro-organisms, the potential of membranes and reuse objectives.

Many examples are given from references in which the author has personally participated at the design and implementation level. The advantages and limitations of the processes are listed.

1.7. Glossary

There are many definitions of treated wastewater reuse in the regulations of many countries.

– *Direct drinking water reuse*: the reused water supply is further treated and then sent directly to a drinking water treatment plant or distribution system, with or without intermediate storage.

– *Direct recharge*: the treated wastewater exiting the wastewater treatment plant is directly injected into the groundwater. Direct recharge is usually accomplished via injection wells or equivalent.

– *Direct reuse*: it consists of not discharging the treated wastewater into a river, but reinjecting it immediately into a network for dedicated use (irrigation, watering of green spaces or golf courses, road cleaning, etc.).

– *Indirect reuse*: the water is taken from a stream or an alluvial aquifer, while at a short distance upstream a wastewater treatment plant has discharged wastewater after treatment. Treated water has been diluted in the receiving environment.

– *Planned reuse*: it includes situations in which treated wastewater is used directly or indirectly under quantitative and qualitative control.

– *Unplanned reuse*: it includes situations where supplies are from environments where wastewater has been discharged, and the quantity and quality of which is not controlled by the user.

– *Wastewater recycling*: often, this term is synonymous with reused water. However, in some countries, recycled water is more closely linked to internal recycling within an industry. The wastewater is collected, treated and then returned to the same industrial process circuit. In Canada, the term recycled water is defined as the return of treated wastewater to be reused for purposes that generate wastewater, for example, the treatment and recycling of all domestic wastewater for flushing toilets and other non-drinking uses.

1.8. References

Anderson, J. (2003). The environmental benefits of water recycling and reuse. *Water Sci. Technol.*, 3(4), 1–10.

Baum, R., Luh, J., Bartram, J. (2013). Sanitation: A global estimate of sewerage connections without treatment and the resulting impact on MDG progress. *Environmental Science & Technology*, 47, 1994–2000.

Bixio, D., De Heyder, B., Cikurel, H., Muston, M., Miska, V., Joksimovic, D., Schäfer, A.U., Ravazzini, A., Aharoni, A., Savic, D., Thoeye, C. (2005). Municipal wastewater reclamation: Where do we stand? An overview of treatment technology and management practice. *Water Science and Technology*, 5(1), 77–85.

Bréda, N. and Badeau, V. (2008). Forest tree responses to extreme drought and some biotic events: Towards a selection according to hazard tolerance? *Compte rendus géosciences*, 340(9/10), 651–662.

BRGM (2010). La réutilisation des eaux usées : un enjeu majeur de développement durable. Report, BRGM.

Cerema (2020). Réutilisation des eaux usées, le panorama français. Report, Cerema.

Crook, J. (1998). Water reclamation and reuse criteria. In *Wastewater Reclamation, Recycling and Reuse*, Asano, T. (ed.). US EPA, New York.

Degrémont, S.A. (2005). *Mémento technique de l'eau*. Degrémont & Suez, Paris.

Dolnicar, S., Hurlimann, A., Grun, B. (2011). What affects public acceptance of recycled and desalinated water? *Water Research*, 45, 933–943.

Duhamel, A., Habets, F., Déqué, M., Barros, V., Tignor, M. (2009). Impact du changement climatique sur les ressources en eau et les extrêmes hydrologiques dans les bassins de la Seine et la Somme. Final Report on the RExHySS Project, GIEC Program.

Faranda, D., Salvatore, P., Bulut, B. (2023). Persistent anticyclonic conditions and climate change exacerbated the exceptional 2022 European-Mediterranean drought. *Environmental Research Letters*, 18(3), 1–12.

Gaid, K. (2023). *Drinking Water Treatment 1*. ISTE Ltd, London, and John Wiley & Sons, New York.

Garnier, E. (2010). Bassesses extraordinaires et grandes chaleurs. Sécheresses et chaleurs en France. *La Houille blanche*, 96(4), 26–42.

Hughes, G., Chinowsky, P., Strzepek, K. (2010). The costs of adaptation to climate change for water infrastructure in OECD countries. *Utilities Policy*, 18, 142–153.

Hughes, J., Cowper-Heays, K., Olesson, E., Rob Bell, R., Stroombergen, A. (2021). Impacts and implications of climate change on wastewater systems: A New Zealand perspective. *Climate Risk Management*, 31, 1–19.

IPCC (2021). Summary for policymakers. In *Climate Change, The Physical Science Basis. Contribution of Working Group I to the Sixth Assessment Report of the Intergovernmental Panel on Climate Change*, Masson-Delmotte, V. (ed.). Cambridge University Press, Cambridge.

Isobe, M. (2013). Impact of global warming on coastal structures in shallow water. *Ocean Engineering*, 71, 51–57.

Langeveld, J.G., Schilperoort, R.P.S., Weijers, S.R. (2013). Climate change and urban wastewater infrastructure: There is more to explore. *Journal of Hydrology*, 476, 112–119.

Loucks, D.P. (2000). Sustainable water resources management. *Water Inter.*, 25(1), 3–10.

Major, D.C., Omojola, A., Dettinger, M., Hanson, R.T., Sanchez-Rodriguez, R. (2011). Climate change, water, and wastewater in cities. In *Climate Change and Cities: First Assessment Report of the Urban Climate Change Research Network*, Rosenzweig, C., Solecki, W.D., Hammer, S.A., Mehrotra, S. (eds). Cambridge University Press, Cambridge.

Mark, O., Svensson, G., König, A., Linde, J.J. (2008). Analyses and adaptation of climate change impacts on urban drainage systems. In *11th International Conference on Urban Drainage*, Edinburgh.

Moisselin, J.-M. and Dubuisson, B. (2006). Évolution des extrêmes de températures et de précipitation en France au XXe siècle. *La Météorologie*, 54, 33–42.

Moisselin, J.-M., Schneider, M., Canellas, C., Mestre, O. (2002). Les changements climatiques en France au XXe siècle. *La Météorologie*, 38, 45–56.

Onema (2022). La réutilisation des eaux usées. Report, Onema.

Peterson, J.D., Murphy, R.R., Jin, Y., Wang, L., Nessl, M.B., Ikehata, K. (2011). Health effects associated with wastewater treatment, reuse, and disposal. *Water Environment Research*, 83, 1853–1875.

PNUD (2009). Problématique du secteur de l'eau et impacts liés au climat en Algérie. Report, Programme des Nations Unies pour le développement (PNUD), Alger.

Ribes, A., Boé, J., Qasmi, S., Dubuisson, B., Douville, H., Terray, L. (2022). An updated assessment of past and future warming over France based on a regional observational constraint. *Earth System Dynamics*, 13(4), 1397–1415.

Russell, S. and Hampton, G. (2006). Challenges in understanding public responses and providing effective public consultation on water reuse. *Desalination*, 187, 215–227.

Shakeri, H., Motiee, H., Mc Bean, E. (2021). Forecasting impacts of climate change on changes of municipal wastewater production in wastewater reuse projects. *Journal of Cleaner Production*, 329, 121–132.

Shilklomanov, I.A. (2000). Appraisal and assessment of world water resources. *Water Inter.*, 25(1), 11–32.

Tolkou, A. and Zouboulis, A. (2015). Effect of climate change in wastewater treatment plants: Reviewing the problems and solutions. In *Managing Water Resources Under Climate Uncertainty*, Shrestha, S., Anal, A.K., Salam, P.A., van der Valk, M. (eds). Springer, Berlin.

Trinh, L.T., Duong, C.C., Van Der Steen, P., Lens, N.L. (2013). Exploring the potential for wastewater reuse in agriculture as a climate change adaptation measure for Can Tho City, Vietnam. *Agricultural Water Management*, 128, 143–154.

2

Characterization of Treated Wastewater

2.1. Main parameters defining the quality of treated wastewater

Treated wastewater contains solids, dissolved substances and micro-organisms. These are the main reason for restrictions on reuse, as they very often contain elements that are harmful to health, organic micro-pollutants containing pathogens such as viruses, bacteria, protozoa and helminths. The risk of water-borne infection by pathogens depends on a set of factors that include their concentration, dispersion in water, the ability of these intestinal agents to survive in the environment, the quality of water treatment, the infective dose, exposure and susceptibility of the exposed population.

The composition of wastewater exiting treatment plants includes various organic substances, biodegradable or not, mineral substances such as phosphates, ammonia, nitrates and nitrites, as well as solids represented by suspended matter. Table 2.1 shows the limits.

They must not contain substances that may promote odors.

The temperature of the effluent released shall be below 30°C unless the upstream temperature exceeds 30°C. In this case, the temperature of the effluent discharged shall not be higher than the temperature of the upstream body of water.

The pH of the effluent discharged shall be between 5.5 and 8.5.

Other constituents such as micro-pollutants, metals and salts are moderately, slightly or not at all eliminated, depending on the type of treatment applied to the plant. It is generally estimated that the same concentrations as those obtained in raw wastewater will be found at the outlet.

Parameter	Concentration		Minimum treatment efficiency compared to input values
BOD5 (biological oxygen demand)	25 mg·L^{-1} O_2	Or	70% for plants treating a load between 120 and 600 kg·day^{-1} of BOD5 80% for plants treating a load >600 kg·day^{-1} of BOD5
COD (chemical oxygen demand)	125 mg·L^{-1} O_2	Or	75%
SS (suspended solids)	35 mg·L^{-1} O_2	Or	90%
	Sensitive area		**Minimum treatment efficiency compared to input values**
NGL (global nitrogen)	15 mg·L^{-1} for a pollution load between 600 and 6,000 kg·day^{-1} 10 mg·L^{-1} for a pollution load >6,000 kg·day^{-1}	Or	70%
TP (total phosphorus)	2 mg·L^{-1} for a pollution load between 600 and 6,000 kg·day^{-1} 1 mg·L^{-1} for a pollution load >6,000 kg·day^{-1}	Or	80%

Table 2.1. *Example of physicochemical parameters of treated wastewater discharge in a receiving environment (as of June 22, 2007 and of July 21, 2015, France)*

The microbiological composition of treated wastewater is extremely variable according to season, origin of the wastewater collected, health status of populations, treatment applied in the wastewater treatment plant, etc. As a result, they contain a wide variety of micro-organisms at varying concentrations, potentially pathogenic to humans (bacteria, moulds and their toxins, viruses and parasites) and likely to have health effects via the respiratory tract, skin, mucosa and/or ingestion depending on the sensitivity of the exposed person and the dose of micro-organisms to which they are exposed (Michael-Kordatou et al. 2015).

The main predominant parameters in treated wastewater that deserve special attention are:

– the indicators of organic matter (biological oxygen demand and chemical oxygen demand);

– suspended solids (SS);

– mineral substances;

– heavy metals;

– microbiology.

Carbonaceous organic matter in biologically treated urban wastewater is composed of a heterogeneous mixture of organic compounds with various structures and varying origins, including natural organic matter, soluble microbial products, pharmaceutical residues, protein compounds, carbohydrates, fats, urea, surfactants, phenols, industrial residues, bacterial degradation by-products, pesticides and endocrine disruptors. This high heterogeneity in their composition must be taken into account when reused for purposes such as direct or indirect drinking water, specific industrial applications, groundwater recharge and irrigation of food eaten raw.

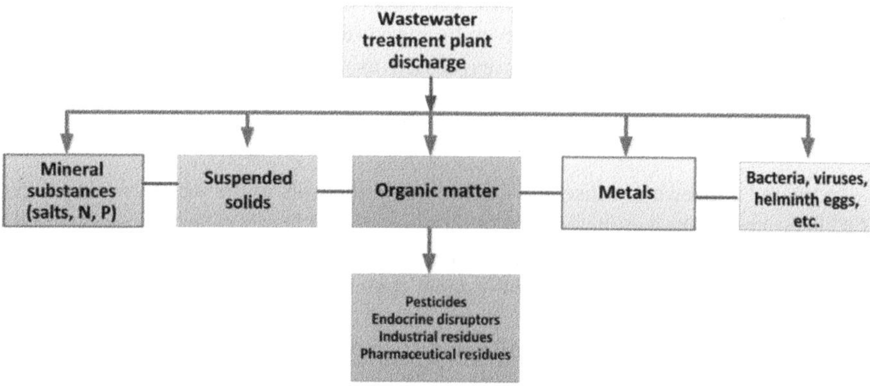

Figure 2.1. *Physicochemical and microbiological characteristics of wastewater*

2.1.1. *Indicators of organic matter: biological oxygen demand, chemical oxygen demand and total organic carbon*

Biological demand in O2 (BOD5) represents the amount of oxygen required to oxidize organic (biodegradable) matter by biological means (oxidation of biodegradable organic matter by bacteria). Biological oxygen demand (BOD) is a water pollution index that allows the evaluation of the biodegradable fraction of carbon pollutant load in wastewater and is generally calculated after 5 days at 20°C and protected from light, which is called BOD5.

BOD is the potential dioxygen consumption biological means. This parameter is a good indicator of the content of biodegradable organic matter in water. BOD is used to measure the quality of water in a sample: surface waters: rivers, lakes, etc.; wastewater: treatment plants, industrial waste, etc. It can be considered as an indirect indication of biodegradable organic carbon.

Chemical oxygen demand (COD): The oxygen amount required to chemically oxidize organic and inorganic oxidizable matter contained in a sample. This parameter gives an estimate of the number of pollutants present in wastewater. It is one of the most widely used methods for assessing the overall organic substance load.

COD measures all oxidable substances, whether biodegradable or not (persistent organic or organometallic pollutants).

For more accuracy with treated wastewater, it is recommended to use *the total organic carbon* and dissolved organic carbon (TOC and DOC), which are much more reliable. Organic carbon can even be measured in-line and in situ. It indicates the total amount of organic matter expressed as carbon ($mg \cdot L^{-1}$) contained in a solution. It makes it possible to know directly the water organic load. DOC is the remaining organic matter after filtration on 0.45-µm membranes.

In treated wastewater reuse and depending on the uses of the water produced, the TOC is an important parameter used to monitor overall levels of organic compounds present. It reflects the importance of having a general indicator that is easy to measure for the approximate level of organic contamination.

When analyzing the TOC, the following are measured:

– TC = total carbon;

– TIC = total inorganic carbon (carbonate-bound);

– POCg = purgeable organic carbon;

– NPOCg = non-purgeable organic carbon;

– DOC = dissolved organic carbon;

– pOCa = particulate organic carbon.

Volatile organic compounds have a low boiling point and can be purified from a solution by injecting gas into a sample.

The summary is as follows:

$$TOC = TC - TIC$$

$$TOC = POCg + NPOCg$$

Generally, the link is used:

$$TOC = DOC + pOCa$$

After membrane filtration during which all pOCa is eliminated, the following is obtained:

$$TOC = DOC$$

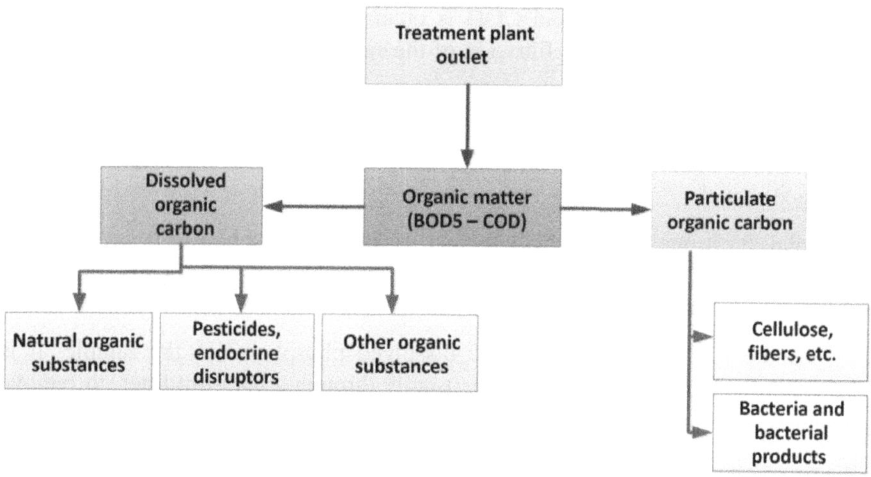

Figure 2.2. *Classification of organic matter in treated wastewater*

This principle is also used after matter filtration when turbidity is <2 NTU. DOC is found in colloids, mainly humic acids, fulvic acids, small molecules from the decomposition of organic matter (proteins, lipids, carbohydrates). COD is the most widely used indicator for design, mathematical modeling and performance evaluation in a design project of a plant intended for treated wastewater reuse.

Analysis of wastewater samples can specify the nature of the pollution.

We distinguish:

– Total BOD: it includes particulate and dissolved organic matter. It is measured without filtering the sample and gives an indication of the total amount of biodegradable organic matter present in the raw sample.

– The particulate or colloidal BOD is obtained by the difference between the gross BOD and the BOD after filtration of the sample. It is mainly composed of SS (MES).

– Soluble BOD: it corresponds to the biodegradable organic matter dissolved in the sample. It is measured after filtration of the sample, usually through a 0.45 μm filter, to remove suspended particles.

– Total COD: it includes particulate and dissolved organic matter. It is measured without filtering the sample and gives an indication of the total amount of organic matter present in the raw sample, whether or not it is biodegradable.

– The particulate or colloidal COD is obtained by the difference between the crude COD and the COD after filtration of the sample. It is mainly composed of SS (MES).

– Soluble COD: it corresponds to the dissolved organic matter in the sample. It is measured after filtration of the sample, usually through a 0.45-μm filter, to remove suspended particles.

– Total P: it includes particulate and dissolved phosphorus. It is measured without filtering the sample and gives an indication of the total amount of phosphates present in the crude sample.

– Soluble P: it corresponds to the dissolved phosphorus in the sample. It is measured after filtration of the sample, usually through a 0.45-μm filter, to remove suspended phosphate particles.

This is as follows:

Total BOD = particulate BOD + soluble BOD

Total COD = particulate COD + soluble COD

Total P = particulate P + soluble P

According to Table 2.1, the COD/BOD5 ratio in treated wastewater is 5, meaning that the inert (non-biodegradable) fraction is high. It indicates that a physicochemical treatment is necessary to improve water quality.

For non-agricultural uses, biological treatment may be used to remove residual nitrogen if it cannot be completely treated at the treatment plant.

2.1.2. *Suspended solids*

SS (or suspended particles) refer to insoluble solids that are visible to the naked eye and are suspended in a liquid. Total suspended solids (TSS) are a measurement of SS.

SS are also studied as a "matrix" capable of adsorbing various pollutants, which can be transported by water in wastewater treatment plants. A distinction is made between:

– suspended colloids (size between 10 nm and 1 μm);

– dissolved matter (size less than 10 nm).

Fine particles suspended in treated wastewater are either of natural origin, in connection with precipitation, or of anthropogenic origin and then brought by atmospheric deposition and/or urban, agricultural, domestic rural and industrial discharges.

SS can be classified according to their size and condition, chemical characteristics and settleability.

In general, dissolved solids are considered to be those with a diameter of less than 10^{-3} μm, colloidal solids are considered to be those with a diameter of between 10^{-3} and 100 μm and SS those with a diameter greater than 100 μm.

If solids are subjected to a high temperature (550°C), the organic fraction is oxidized (volatilized) leaving after combustion only the inert (non-oxidized) fraction. Volatile solids represent an estimate of organic matter in solids, while non-volatile (fixed) solids represent inorganic or mineral matter.

Settleable solids are considered to be capable of settling within a 1-h period. The volume of solids accumulated at the bottom of a container called an Imhoff cone is measured and expressed in $mL \cdot L^{-1}$. The fraction that does not settle represents the non-settleable solids.

SS are mostly biodegradable. The majority of pathogenic micro-organisms in wastewater are transported by SS. In treated water, the SS concentration limit of 35 $mg \cdot L^{-1}$ requires a much more advanced treatment to cope with health risks and applied technologies. For example, excessive SS can lead to difficulties in the transport and distribution of effluents, as well as the clogging of irrigation systems.

2.1.3. *Nitrogen*

Several forms of nitrogen are present at the outlet of the treatment plants:

– oxidized nitrogen: nitrites and nitrates;

– non-oxidized forms: Kjeldahl nitrogen, including organic nitrogen and ammoniacal nitrogen (NH_4^+);

– ammonifiable organic nitrogen;

– refractory organic nitrogen.

According to the regulations of the countries, the residual form of nitrogen can be ammoniacal if discharge standards do not impose a limit value for this parameter. Consequently, its presence in organic or ammoniacal form results in oxygen consumption in the natural environment and alters the living conditions of the environment.

In France, where standards for global nitrogen are sufficiently strict, nitrification and denitrification must be controlled in an optimal way.

Nitrogen form	Formula	Oxidation state
Molecular nitrogen	N_2	0
Organic nitrogen	Variable	Variable
Ammonia (gas)	NH_3	–3
Ammonium ion	NH_4^+	–3
Nitrite ion	NO_2^-	+3
Nitrate ion	NO_3^-	+5

Table 2.2. *Nitrogen predominant forms in treated wastewater*

The overall nitrogen includes Kjeldahl nitrogen (TKN), nitrites and nitrates.

Namely:

$$TN = TKN + \text{Nitrites } (NO_2^-) + \text{Nitrates } (NO_3^-)$$

$$TKN = \text{Ammoniacal nitrogen } (NH_4^+) + \text{Organic nitrogen}$$

The form of ammonia nitrogen depends on the pH:

– at pH < 8, ammonia is in the form of NH_4^+;

– at pH 9.5, overall 50% NH_3 and 50% NH_4^+ are obtained;

– at pH > 11, almost all the ammonia is in NH_3 form.

The following formula allows to calculate the proportion of free ammonia in comparison to total ammonia present in water, depending on temperature and pH (Emerson et al. 1975):

$$\frac{\text{Free NH}_3}{\text{Total NH}_3}\% = \left(1 + 10^{0.09018+\left(\frac{2{,}729.92}{T+273}\right)-\text{pH}}\right)^{-1} \times 100$$

with T: temperature (°C).

In the aquatic environment, nitrogen can be found in the forms shown in Table 2.2. Nitrogen is an important component in the water cycle.

In treated wastewater reuse, this parameter should be given special attention according to the application. Nitrogen is an essential nutrient for algae, which under certain conditions leads to eutrophication of lakes and reservoirs. In the event that the direct use project relates to recreational, lake-feeding or reservoir water, precautions will be taken with regard to the development of algae.

In addition, it will result in consumption of dissolved oxygen in the water body due to the processes of converting ammonia into nitrites and nitrates.

2.1.4. *Phosphorus*

Total phosphorus is present as phosphates, according to the distribution:

– inorganic (polyphosphates and orthophosphates) – from detergents and other household chemicals;

– organic (linked to organic compounds) – physiological origin.

Phosphorus in detergents is present in wastewater as soluble polyphosphates or, after hydrolysis, as orthophosphate. These are directly available for biological metabolism.

The forms in which orthophosphates are present in water depend on the pH and include PO_4^{3-}, HPO_4^{2-}, $H_2PO_4^-$, H_3PO_4.

In wastewater, the predominant form is HPO_4^{2-}.

Polyphosphates are more complex molecules, comprising two to several phosphorus atoms. They are converted into orthophosphates by hydrolysis.

Phosphorus in detergents can account for up to 50% of the total phosphorus in wastewater.

Total phosphorus therefore includes:

– soluble (mainly inorganic) phosphorus – mainly polyphosphates and orthophosphates, as well as a small fraction corresponding to the phosphorus bound to the soluble organic matter;

– particulate phosphorus – bound to particulate organic matter.

In the case of treated wastewater reuse, the importance of this parameter is associated with the following problem: phosphorus is an essential nutrient for algae growth, which under certain conditions may lead to eutrophication of lakes and reservoirs. It is estimated that up of 20 $\mu g \cdot L^{-1}$, the effect of P on algae development promotes their growth.

2.1.5. *Pesticides and endocrine disruptors*

After discharge, the treated wastewater is diluted in surface water. This resource is treated to eliminate undesirable compounds for the production of drinking water. The operation of water treatment plants is more or less sophisticated and many processes can be used with different reduction results depending on substances. In treated wastewater reuse, it will be necessary for certain applications to ensure a significant reduction of organic and minerals compounds by using the appropriate technologies.

Over the past decade, a large number of scientific papers have described the performance of wastewater treatment plants with respect to these substances. It appears that the treated waters are still provided with a concentration often not negligible, which is expected, as the equipment available to wastewater treatment plants is not suitable to eliminate these substances.

Main families	Compounds
Drug molecules	Anxiolytics, analgesics, anti-inflammatories, antibiotics, lipid-lowering, antihypertensives, antiepileptics, anti-cancer, iodinated contrast agents, etc.
Cosmetic and personal hygiene products	Musk (galaxolide, tonalide), triclosan, etc.
Endocrine disruptors	Hormones, alkylphenols, BPA, phthalates, flame retardants, etc.
Illegal drugs	Cocaine and metabolites, THC, amphetamines, etc.
Pesticides	Insecticides, herbicides
Priority substances (Regulation (EU) No. 2020/741[1]) and dangerous substances (circular RSDE 050109)	HAP, dichloromethane
Disinfection by-products	Trihalomethane, haloacetic acids

Table 2.3. *Main families and organic compounds*

The most represented families of pesticides include:

– organophosphorus (chlorpyriphos and malathion), which have an insecticidal effect by blocking acetylcholinesterase (disturbance of nerve impulse transmission);

– organochlorines (DDT, chlordecone and lindane), insecticides that affect the nervous system by disrupting nerve impulses at the synapse level. These compounds are all now banned in many countries, but are very persistent in the environment;

– carbamates (aldicarb and carbofuran), which act on the nervous system in the same way as organophosphates;

– triazines (atrazine and simazine), which have a herbicidal effect by blocking the transport of electrons and light energy;

– phenylurea herbicides (diuron and isoproturon), which have a mode of action comparable to that of triazines;

1 Annexes available at: http://www.iste.co.uk/gaid/annexes1.zip. The texts are part of the regulations governing treated wastewater reuse in France. Regulations now actually allow the reuse of treated wastewater from wastewater treatment plants, non-collective sanitation systems and classified environmental protection facilities for uses other than irrigation. It sets out the minimum conditions and requirements for water reuse. These texts set up an authorization procedure so that new uses, such as road cleaning or garden watering, can be implemented.

– chloroacetamide herbicides (metolachlor and acetochlor), which act on plants by inhibiting the elongases, enzymes involved in lipid synthesis (long-chain fatty acids);

– aryloxyacids (2.4 D, mecoprop and dichlorprop), which disrupt the auxin IIA (indole 3-acetic acid) regulation, a phytohormone essential for plant growth;

– sulfonylureas (chlorsulfuron and nicosulfuron), which affect the synthesis of branched amino acids (valine, leucine and isoleucine) through inhibition of acetolactate synthase.

Agricultural activity is not the only source of pesticide emissions. Among the secondary sources are the maintenance of public spaces, golf courses, private use, annual wetland mosquito-control, veterinary sector, road maintenance, etc.

2.1.5.1. *Hormones*

Natural (estradiol and its derivatives) and synthetic (ethinylestradiol from the contraceptive pill) animal hormones are excreted by humans and are found in wastewater. Natural oestrogen hormones include estrone (E1), 17-α-estradiol (E2) and estriol (E3). The most important synthetic hormone, as it is used in birth control pills, is 17-α-ethinylestradiol (EE2). Women of reproductive age excrete at least 50 $\mu g \cdot D^{-1}$ of oestrogen hormones in their urine. This rate can exceed 5 $mg \cdot D^{-1}$ for a pregnant woman. Men also excrete hormones, but in much smaller proportions, approximately a few micrograms. Testosterone is also excreted but is very well eliminated by mineralization in wastewater treatment plants (Barbier 2011).

Concentrations are therefore very variable. However, reduction rates are significant, in the range of 70–85%, except for ethinylestradiol, where it varies from half to two-thirds.

Other products are likely to have such effects as alkylphenols ethoxylates (APEO), which are contained in industrial cleaning products, paints and some coatings. Among these products, octylphenol and nonylphenol have an endocrine disrupting effect that has been found up to concentrations as high as 10 $\mu g \cdot L^{-1}$. Phthalates such as DEHP (di(2-ethylhexyl) phthalate) and BBP (benzyl butyl phthalate) range from 0.2 to 1 $\mu g \cdot L^{-1}$.

Drug residues are among the main xenobiotics present at the plant outlet and in the environment. Not all of them are endocrine disruptors.

The most common products are carbamazepine, an anticonvulsant and lopromides, which are contrast agents for radiology. Concentrations of the substances sought are in the range of 10–100 $ng \cdot L^{-1}$ in surface waters but are much higher at the outlet of the treatment plants, approximately 1 $\mu g \cdot L^{-1}$.

	Plant inlet		Plant outlet	
	Min.	Max.	Min.	Max.
		A		
Estradiol	0.61	51.7	0.04	19.1
Estrone	0.90	17.6	0.04	3.62
Estriol	1.11	14.9	0.22	3.14
Ethinyl E2	1.03	48.49	0.3	14.89
		B		
Estradiol	0.4	14.41	0.17	2.2
Estrone	0.13	19.8	0.06	7.75
Estriol	0.45	18	0.09	3.58
Ethinyl E2	0.77	4.09	0.3	1.47
		C		
Estradiol	0.25	36.2	0.003	2.69
Estrone	2.19	13.2	0.2	6.09
Estriol	0.8	9.75	0.17	3.65
Ethinyl E2	1.07	8.89	0.46	2.93

Table 2.4. *Concentrations in estrogens at the inlet and outlet of a few wastewater treatment plants*

Substance	**Family**	**Performance (%)**	**Wastewater – inlet (µg·L^{-1})**	**Wastewater treated outlet (µg·L^{-1})**
4-Nonylphenol	Alkylphenols	84 (60–100)	15.7	1.3
Atrazine	Pesticide	2 (1–30)	0.02	0.02
4-t-Octylphenol	Alkylphenols	88 (70–100)	5.6	0.21
Bisphenol A	Phenol polymers	60 (30–90)	0.16	0.06
4-NP1EO	Alkylphenols	88 (70–100)	9.0	0.47
DEHP	Phthalate	92 (80–100)	52.8	4

Table 2.5. *Performance of wastewater treatment plants for a few endocrine disruptors (Lopez and Laurent 2013)*

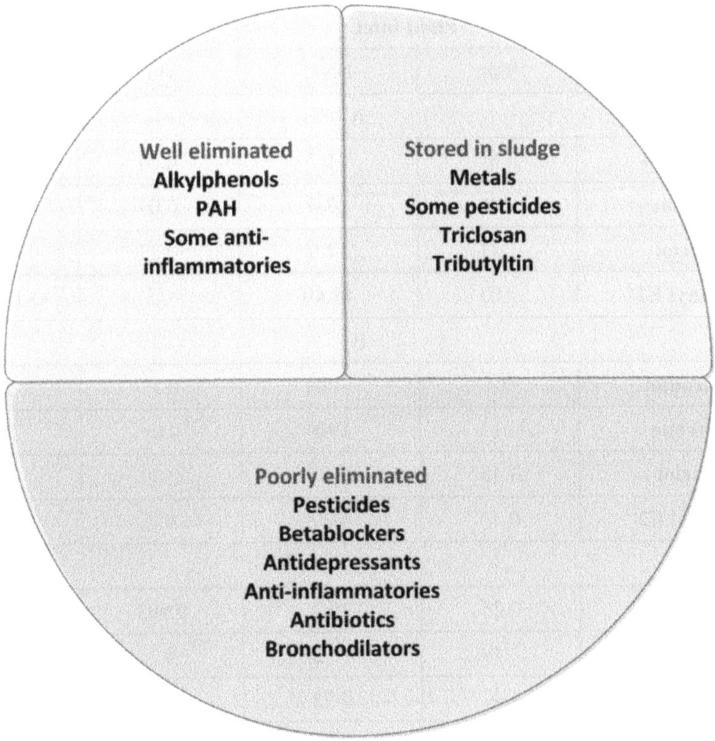

Figure 2.3. *The fate of a few substances in wastewater treatment plants*

Finally, antibiotics are also a concern as they are present in all aquatic compartments and accumulate in sediments.

– *Flame retardant polybrominated compounds*: the most common are polybrominated diphenyl ethers (PBDEs), different types of hexabromocyclododecane (HBCD) and tetrabromobisphenol A (TBBPA). These are *persistent organic lipophilic and bioaccumulative* substances.

– *Perfluorinated compounds*: the two compounds that are most in focus are perfluorooctane sulfonate (PFOS) and perfluorooctanoic acid (PFOA). They are hydrophobic and lipophobic.

– *Parabens*: due to their ubiquity, parabens are found in almost all wastewaters.

In France, more than 232 substances identified as pesticides, endocrine disruptors, drug residues and industrial residues were found in urban wastewater.

In summary, based on the multiple analyses carried out at the wastewater plant discharges, we know how to understand the fate of these substances when they pass through the wastewater treatment plants.

Pesticides, drug residues and endocrine disruptors are still not specified in guidelines of the discharge of treated wastewater, and they are therefore present in the receiving environment.

2.2. Microbiological aspects

Organism	Description
Bacteria	Unicellular organism. Present in different forms and different sizes. Responsible for the degradation of organic matter and nitrogen compounds in treatment plants. Responsible for water-borne or food-borne disease. *E. coli* (enterotoxin): *Leptospira*, *Salmonella*, *Shigella*, *Yersinia enterocolitica*, *Pseudomonas aeruginosa* (pseudomonas aeruginosa), *Staphylococcus*, *Streptococcus*, *Klebsiella pneumonia.* Diarrhea: fever, epidermal hepatitis, meningitis; gastroenteritis, vomiting; colitis, colic, diarrhea; fever, enteritis, erythema; urinary tract inflammation; abscess, respiratory tract inflammation; local infection, endocarditis, otitis; pneumonia.
Viruses	Pathogenic and difficult to eliminate during water treatment. Include enteroviruses, hepatitis A virus, Norwalk-type gastroenteritis virus, reovirus and rotavirus, adenovirus. Wastewater is one of the sources of contamination.
Enteric protozoa	Some enteric protozoa are pathogenic and have been associated with outbreaks of diseases transmitted by drinking water. The most worrying protozoa are *Giardia* and *Cryptosporidium*. These organisms can be found in water as a result of direct or indirect contamination by human or animal feces. *Giardia* and *Cryptosporidium* are often transmitted from person to person.
Helminths	Helminths, parasitic worms of the digestive tract of humans or animals, release in large numbers into the outer environment eggs with a shell composed of several membranes more or less thick. This resistant structure allows the biological cycle to be completed and the infestation power of the worm to continue. These microscopic eggs have a size varying between 30 and 160 µm. Intestinal helminthiasis are parasitic diseases of animals and humans caused by the infestation of worms, helminths. These pathogens can be transmitted to humans through direct contact with wastewater, or indirectly through the consumption of crops irrigated with such wastewater, or by food of animal origin.

Table 2.6. *Most common microorganisms found in treated wastewater*

Treated wastewater still contains a large proportion of the pathogenic microorganisms excreted with fecal matter. Most of these organisms play different essential roles, mainly related to the transformation of constituents in biogeochemical cycles. The biological treatment of wastewater is based on the activities of these microorganisms. Another important aspect of the biological quality of water is that it is related to disease transmission by pathogenic organisms. Microorganisms are the cause of the main restrictions on wastewater direct reuse.

Several pathogenic microorganisms are found in domestic wastewater and can cause different diseases. They can be grouped into four categories: bacteria, viruses, protozoa and helminths. Among the most important human pathogenic bacteria that can be transmitted directly (consumption, bathing, aerosols) or indirectly (contact with food or contaminated surface) by water, *Salmonella, Shigella, Vibrio cholerae,* enterohemorrhagic strain of *Escherichia coli* (O157:H7), *Yersinia enterocolitica, Campylobacter jejuni, Leptospira* and mycobacteria. These pathogenic microorganisms can cause gastroenteritis and other more serious conditions (Table 2.6).

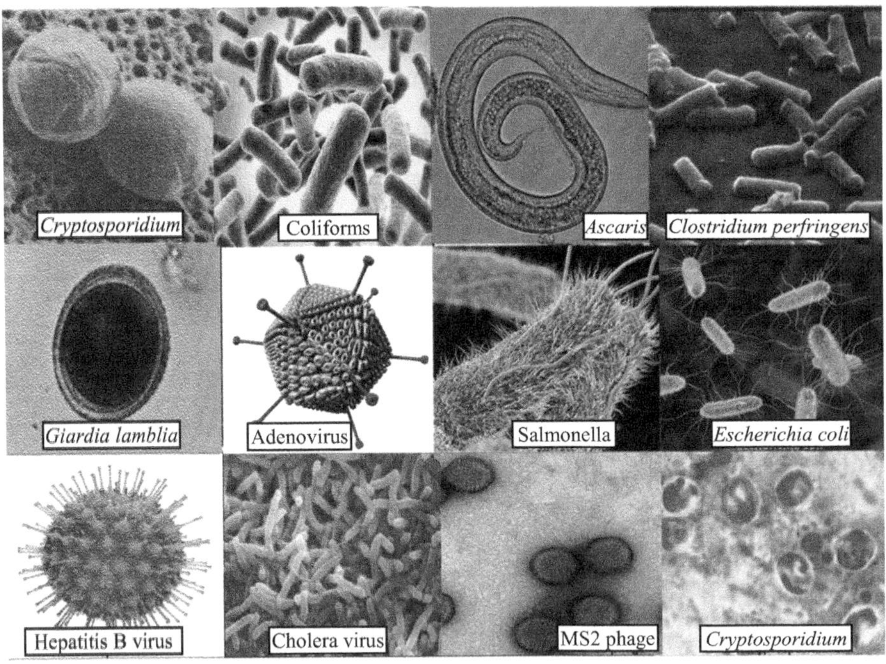

Figure 2.4. *A few microorganisms found in wastewater at the outlet of wastewater treatment plants. For a color version of this figure, see www.iste.co.uk/gaid/reuse1.zip*

2.2.1. *Bacteria*

2.2.1.1. *Coliforms*

Coliforms have good resistance in water compared to other enteric pathogenic bacteria. This characteristic is important, and that is one of the reasons why they are a good indicator of fecal contamination. The group of coliform bacteria mainly includes species of the genera *Citrobacter, Enterobacter, Escherichia* and *Klebsiella* and includes fecal coliforms (FC), of which *Escherichia coli* is the predominant species. Many of the coliforms are capable of growing outside the intestine, especially in hot climates, where their count is not suitable as a monitoring parameter for treated wastewater reuse systems. The fecal coliform test may also include some non-fecal organisms that can grow at 44°C, so that the number of *E. coli* is the most satisfactory indicator parameter for agricultural wastewater application.

The mechanisms for eliminating coliforms from wastewater treatment plants are similar to those used for pathogenic bacteria. Therefore, the elimination of pathogenic bacteria is usually associated with the elimination of coliforms. However, other pathogenic organisms (such as protozoa cysts and helminth eggs) can be eliminated by different mechanisms.

2.2.1.2. *Fecal streptococci*

The group of fecal streptococci includes two main genera: *Enterococcus* and *Streptococcus*. The genus *Enterococcus* includes many species, most of them of fecal human origin; however, some species are of animal origin. All *Enterococcus* have a high tolerance to adverse environmental conditions. Fecal streptococci rarely multiply in polluted waters and are more resistant than *E. coli* and coliform bacteria. For these reasons, they are sometimes used as indicators for bathing water. In treated wastewater reuse, for this application, a number of references exist with these bacteria.

The relationship between the values of FC and fecal streptococci (FC/FS ratio) is used in some countries to give an indication on the origin of contamination, whether it is predominantly human or animal. High FC/FS values would suggest a predominantly human contamination, while low FC/FS values would suggest a predominantly animal contamination. These are special cases, although these connections are not applicable in all cases, and may give erroneous indications of contamination in various reservoirs. This group of organisms includes species mainly associated with animals (*Streptococcus bovis* and *S. equinus*), other species having a wider distribution (e.g. *S. faecalis* and *S. faecium*, present both in humans and other animals), as well as two biotypes (*S. faecalis* var. *liquefaciens* and a typical starch-hydrolyzing *S. faecalis*) which appear to be ubiquitous, occurring in both polluted and unpolluted environments. Counting fecal streptococci in effluents is a simple routine procedure but has the following limitations: the possible presence of non-fecal biotypes in the natural microflora of crops may impair their usefulness in assessing the bacterial quality of crops irrigated with wastewater, and the survival of fecal streptococci at high temperatures is lower than at low temperatures. Additional studies are justified with respect to the use of fecal streptococci as an indicator in tropical conditions and especially to compare survival with that of salmonella (Huot et al. 2008).

2.2.1.3. *Sulfite-reducing Clostridium*

Clostridium perfringens is a species present in lower concentration than *E. coli.* These bacteria are an exclusively fecal sporulating anaerobes normally used to detect intermittent or prior water pollution, due to the long-term survival of its spores. Although this prolonged survival is generally considered a disadvantage for normal purposes, it can be very useful in treated wastewater reuse studies, because *Clostridium perfringens* can have survival characteristics similar to those of viruses or even helminth eggs.

Clostridium spores can survive in water for much longer than organisms from the coliform group and are more resistant to disinfection. Their presence in the disinfected waters may indicate failures in treatment.

The concentration of pathogenic bacteria in treated wastewater is between 10^6 and 10^7 bacteria per 100 mL. The concentration of enterobacteria is approximately 10^5 to 10^6 u·100 mL^{-1}, 10^3–10^4 for streptococci and 10^3–10^4 u·100 mL^{-1} for *Clostridium perfringens*.

2.2.1.4. *Salmonella* spp.

Several species of *Salmonella* may be present in the raw wastewater of an urban community in a developing tropical country, including *S. typhi* (typhoid causative agent) and many others. It is estimated that a count of 7,000 *Salmonella* per liter is typical in a tropical urban sewer with a similar number of *Shigella*, and perhaps 1,000 *Vibrio cholera* per liter. *Shigella* spp. and *V. cholera* are both eliminated more quickly in the environment, so if elimination of salmonella can be achieved, then the majority of other bacterial pathogens will also have been eliminated (Doran et al. 2005).

2.2.2. *Enteric viruses*

2.2.2.1. *Enterovirus*

Approximately 140 types of enteric viruses have been described and are potentially present in wastewater.

All viruses are pathogenic to humans and cause a wide spectrum of diseases and symptoms. Enteroviruses such as poliovirus, reovirus, rotavirus, adenovirus, astrovirus, calicivirus like the Norwalk agent, and A and E hepatitis viruses. Enteric viruses are present in fewer numbers than bacteria in wastewater and do not multiply in this environment. However, they usually have a low infective dose and can survive several months in water under certain conditions.

The concentration of enteroviruses in treated wastewater is between 10^3 and 10^4 u·L^{-1}.

Enteric viruses are the most likely human pathogens to contaminate groundwater (irrigation, groundwater recharge), especially since their extremely small size (20–100 nm) allows them to seep into the ground, eventually reaching aquifers. Depending on the physicochemistry and specific factors of the virus (e.g. size and isoelectric point), viruses can move around easily in the ground and remain for several months in the soils and groundwater.

Enteroviruses have also been shown to be more resistant to disinfection than bacterial indicators.

They give rise to serious diseases, such as polio and meningitis, or a range of minor diseases, such as respiratory infections. Although there is no strong epidemiological evidence of the spread of these diseases through irrigation systems with wastewater, there is a certain risk, and it is desirable to know the extent to which viruses are eliminated by new and existing treatment processes, especially in tropical conditions. Virus counts can only be carried out in a dedicated laboratory as the cell culture techniques required are highly sensitive to bacterial and fungal contamination.

Pathogenic protozoa and enteric viruses are considered priority microbial contaminants because of their important role in epidemics of water-borne diseases and associated public health risks, their long-term survival in the environment, their resistance to conventional water disinfection processes and the often weak or absent correlation with the traditional water quality bacterial indicator.

2.2.2.2. *Rotavirus*

These viruses are known to cause gastrointestinal problems and, although they are generally present in fewer numbers than enteroviruses in wastewater, they are known to be more persistent, so it is necessary to establish their survival characteristics in relation to enteroviruses and indicator organisms in wastewater. It has been argued that the elimination of viruses in wastewater treatment occurs in parallel with the elimination of SS, as most viral particles are associated with solids. Therefore, measurement of SS in treated effluents should be performed routinely.

2.2.2.3. *Bacteriophages*

Bacteriophages may be representative of the presence of viruses, due to their similarities with human enteric viruses. Bacteriophages are specific viruses that infect bacteria, for example, coliphages, which infect *E. coli*. Coliphages can be abundant in wastewater, due to their rapid reproduction rate resulting from the aggression of bacterial cells. They can be representative as indicators of wastewater contamination, due to their greater persistence compared to bacterial indicators, and also as additional indicators of the efficiency of groundwater treatment or protection.

2.2.3. *Enteric protozoa*

Enteric protozoa are found in wastewater in forms that are highly resistant to environmental conditions and disinfection methods. Like viruses, they do not reproduce outside their host and have a low infective dose in comparison with bacteria. The protozoa *Cryptosporidium* and *Giardia* are among the main agents responsible for diarrheal diseases in humans. The protozoa that most often infect humans are *Giardia lamblia*, *Entamoeba histolytica*, *Cryptosporidium*, *Cyclospora*, *Microsporidia* and others. Their resistant forms (cysts of *Giardia* 8–13 μm and oocysts of *Cryptosporidium* 4–6 μm) are excreted in large numbers by infected animals or humans and are omnipresent in surface water. They resist difficult environmental conditions and chemical disinfectants at concentrations commonly used in water treatment plants to reduce bacterial contamination. Several studies have revealed little or no correlation between the fecal bacterial indicator and the concentration of protozoa. These observations highlight the need for regular monitoring of reused water and the implementation of technologies adapted to the reduction of these microorganisms for effective user protection.

2.2.4. *Helminth eggs*

Helminths do not replace bacterial indicators. However, eggs of nematodes, such as *Ascaris*, *Trichuris*, and *Necator americanus*, can be used as indicators of other helminths (cestodes, trematodes and other nematodes), which are eliminated in water and wastewater treatment with the same mechanism (e.g. sedimentation); they are thus indicators of treatment efficiency. Helminths nematodes such as *Ascaris lumbricoides* and *Trichuris trichiura*, flat worms, cestodes such as *Tænia saginata* and *T. solium,* and trematodes such as *Schistosoma mansoni* are also potential pathogenic microorganisms for humans found in wastewater. The effect of exposure to the pathogenic microorganisms mentioned here depends on several factors related to the host (susceptibility and degree of immunity), the pathogenic microorganism and the environment.

Helminth eggs are an important parameter in the assessment of the use of treated wastewater for irrigation, in which workers can be in direct contact with contaminated water and consumers can consume raw or unpeeled vegetables.

Depending on the country and the wastewater collection system, there are between 10 and 100 eggs L^{-1} in the Middle East. Although 1,000 eggs per liter were observed at certain times of the year. Among the helminths, eggs of *Ascaris* and Taenia are predominant. In France, there are no more helminths since the wastewater from slaughterhouses is no longer connected to the urban sanitation network.

Table 2.7 shows the average bacteriological concentrations at the outlet of wastewater treatment plants (conventional activated sludge system).

Microorganism	**Concentration ($u \cdot 100\ mL^{-1}$)**
Total coliforms	10^6–10^7
Fecal coliforms	10^5–10^6
Streptococci	10^3–10^4
Enterococci	10^3–10^4
Clostridium perfringens	10^3–10^4
Enteroviruses	10^2–10^3
Oocysts of *Cryptosporidium parvum*	10^2–10^4
Cysts of *Giardia lamblia*	10^2–10^4
Helminth eggs	0–10^2

Table 2.7. *Average bacteriological concentrations at the discharge of wastewater treatment plants (conventional activated sludge system)*

These values are average concentrations observed at different wastewater treatment plants around the world. However, they may vary according to the seasons, especially when the pollutant flow is greater, during the periods of slaughter of animals, or when the flow rate is lower while the pollutant load is higher, etc. It should be noted that a reduction of these parameters will be necessary using physical barriers (membranes) or strong chemical oxidants.

2.2.5. *Fecal contamination indicators*

2.2.5.1. *Assessment of treated wastewater*

Given the wide variety of pathogenic microorganisms in water and the limitations of detection methods, acceptable levels in treated wastewater are established for microorganisms that indicate fecal or treatment contamination.

Important criteria to be considered when selecting an indicator microorganism are:

– the indicator organism for fecal pollution must be present only if there is fecal contamination and it must not reproduce outside its host;

– the treatment survival of the indicator microorganism in the environment must be equal to or greater than that of the target microorganism;

– detection and enumeration should be quick and inexpensive;

– the indicator microorganism must be non-pathogenic to humans. No indicator available at this time meets all these criteria, so different microorganisms and parameters are used as indicators for specific purposes (Huot et al. 2008; Santé Canada 2010).

Indicator organisms are used as indicators of the efficiency of pathogen elimination in the wastewater treatment process. The pathogenic organisms that can be represented are bacteria and viruses, since they are eliminated by the same mechanisms as coliform bacteria. Protozoa cysts and helminth eggs, which are mainly eliminated by physical mechanisms, such as sedimentation and filtration, are not well represented by coliform bacteria as indicators of treatment efficiency. There are various other indicator organisms, each with its advantages and disadvantages.

The most commonly used indicators of fecal contamination are total coliforms (TC), FC or thermotolerant coliforms, *Escherichia coli*. FC are a group of bacteria primarily found in the intestinal tract of humans and other animals. This group includes the genus *Escherichia* and, to a lesser extent, the species of *Klebsiella*, *Enterobacter* and *Citrobacter*.

For decades, fecal bacteria (e.g. FC and *E. coli*) have been used to identify fecal contamination and indicate the probable presence of microbial pathogens in water. However, their concentrations are rarely well correlated with those of the pathogens. Thus, bacterial indicators can indicate the probable presence of pathogens in water but cannot accurately predict their level of occurrence. Compared to viruses and protozoa, bacterial indicators are less persistent in aquatic environment and less resistant to disinfection mechanisms and other treatment processes.

There is no quantifiable relationship between TC and pathogenic microorganisms. However, their presence may mean the presence of other pathogenic bacteria. In direct reuse, their presence would mean that the treatment is not sufficient and that the water may be contaminated by other pathogenic microorganisms. They would thus be used as indicators of the water treatment efficiency.

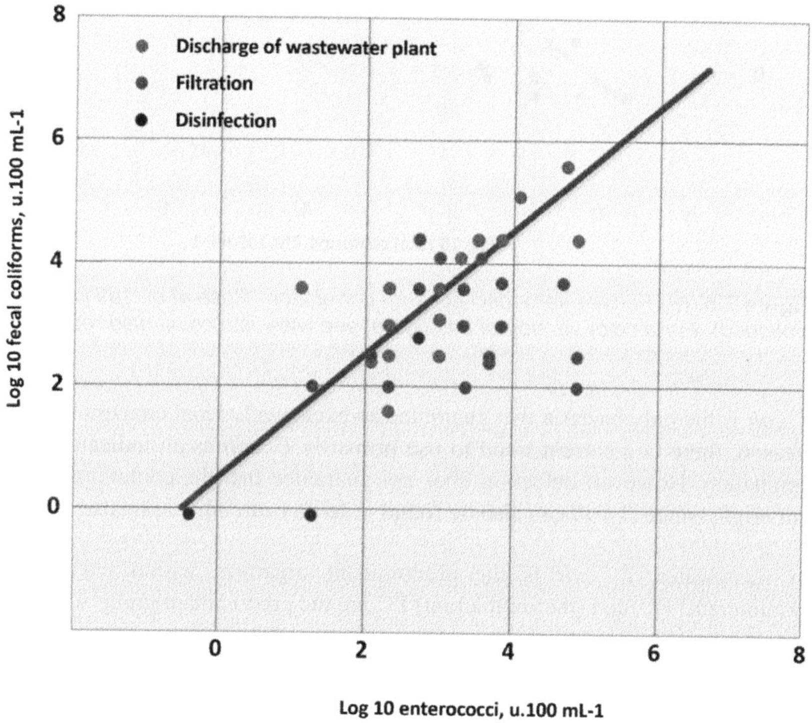

Figure 2.5. *(A) Fecal/enterococci coliform correlation (Rose et al. 1991, 1996). For a color version of this figure, see www.iste.co.uk/gaid/reuse1.zip*

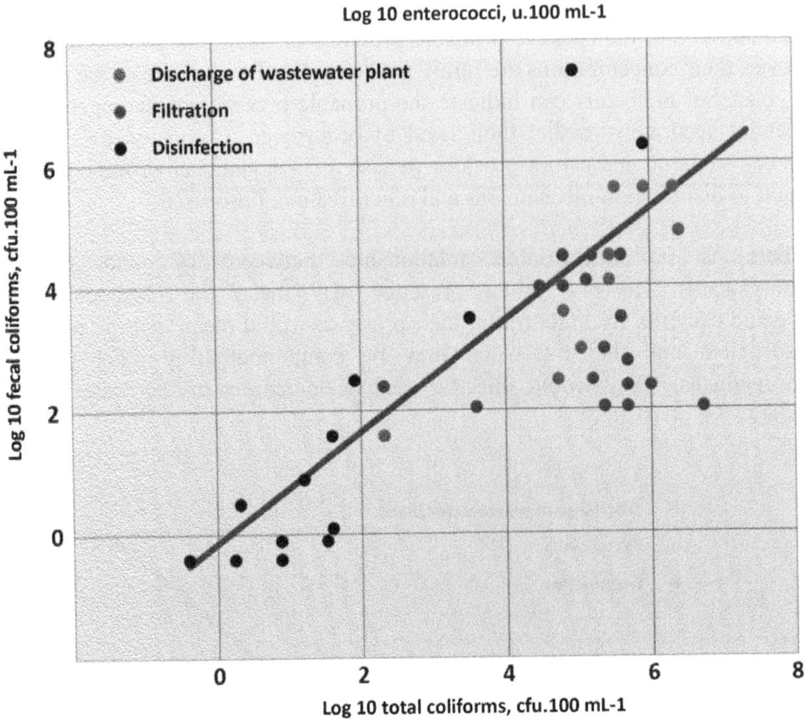

Figure 2.5. *(B) Fecal/enterococci coliform correlation (Rose et al. 1991, 1996) (continued). For a color version of this figure, see www.iste.co.uk/gaid/reuse1.zip*

E. coli is the only bacteria that guarantee an exclusively fecal contamination. For this reason, there is a current trend to use primarily *E. coli* as an indicator of fecal contamination. However, detection does not guarantee that the contamination is of human origin, since *E. coli* can also be found in feces from other animals.

In wastewater, *E. coli* is the predominant organism within the group of (thermotolerant) FC, and (thermotolerant) FC are the predominant group within TC.

Up to now, the most widely used indicator microorganisms for fecal pollution and treatment are coliforms. Coliform limits to be respected are included in most treated wastewater reuse guidelines. The presence of TC would be a very imprecise risk indicator. In fact, these bacteria can grow in a water system, whose treatment plant is perfectly functional and pathogenic microorganisms may be present in distributed water in the absence of TC. However, they are very useful as indicators of the reused water treatment efficiency according to use (drinking water, water in direct contact with the user). Fecal or thermotolerant coliforms can indicate the intrusion of fecal matter, and therefore pathogenic microorganisms, into distribution networks. As some FC are not necessarily of fecal origin, the use of *Escherichia coli* is increasingly recommended, which is the only fecal coliform that is unequivocally always of fecal origin. Its detection must give serious cause for suspicion of fecal pollution and potentially pathogenic microorganisms.

Coliforms are not good indicators of the presence of viruses or parasites (protozoa and helminths) that would be more persistent in the environment (50% reduction time in surface water concentration: coliforms = 0.9 day, *Cryptosporidium* = up to 150 days), more resistant to certain treatments such as disinfection and have lower infective doses. To this end, other indicators such as coliphages (for viruses), *Clostridium perfringens* (for viruses and parasites), enterococci (for viruses and parasites) and a physicochemical parameter such as turbidity (for parasites) are recommended. Some countries or organizations have added, in addition to the coliform criteria, criteria to be respected for parasites and viruses in reused wastewater. Despite their limitations, FC and *E. coli* have several advantages and are the most widely used indicator microorganisms at present.

The current scientific consensus is that there is no universal indicator of water quality, especially for treated and reused wastewater. Each situation is specific, and it is important to select the best pathogenic indicator group(s).

– As process indicators, phage groups are often successfully used as enterovirus substitutes in the evaluation of the efficiency of water treatment processes. This is the case for filtration and disinfection (Rose et al. 1991, 1996).

– As tracers, bacteriophages are often used both in the field and in the laboratory as biocolloids to estimate the fate and transport of pathogenic viruses on the surface and in groundwater. This use of phages as pathogen transport substitutes applies to the assessment of the potential health risk of pathogens in groundwater and the assessment of treatment process efficiency to reduce pathogens in water.

Microorganism	Pollution indicator
Escherichia coli	Fecal pollution indicator; potential presence of enteric pathogens
Enterococci	Fecal pollution indicators; indirect probability of presence of enteric viruses
F-RNA phages	Contamination indicators by enteric viruses Indicators as virus substitutes, to teste a process efficiency
Phages of *B. fragilis*	Fecal contamination indicators. Used as a monitoring parameter of microbial sources
Clostridium perfringens	Fecal pollution indicator. Indicator as virus and protozoa (cysts and oocysts) substitutes to test a disinfection process efficiency

Table 2.8. *Examples of contamination indicators*

Table 2.8 shows the values obtained at the outlet of wastewater treatment plants in France, where discharges are made in receiving environments located in sensitive areas.

Quality parameter $mg \cdot L^{-1}$	Treatment plant outlet (sensitive area)	
Biochemical demand in O_2, BOD5	15	9
Chemical demand in O_2, COD	55	39
Suspended solids	19	12
$N\text{-}NH_4^+$	3	2
$N\text{-}NO_3^-$	8	4
$P\text{-}PO_4^{3-}$	1.5	2
Ca^{2+}	85	71
K^+	15	18
SO_4^{2-}	51	39
Cl^-	38	31
Total dissolved solids	598	450
Boron	0.51	0.28
Total coliforms ($u \cdot 100\ mL^{-1}$)	10^6	10^5
Fecal coliforms ($u \cdot 100\ mL^{-1}$)	10^5	10^4
Helminths ($u \cdot L^{-1}$)	0	0

Table 2.9. *Data from a few plants on the quality of treated wastewater in France*

2.2.5.2. *Current concerns*

Any wastewater can be treated to a very high performance to make drinking water that meets standards guidelines.

However, for a wastewater project, the questions that arise for each project are as follows:

– Which components need to be eliminated?

– How far should the detection of organic substances be pushed?

– From which quality can water be considered as acceptable for drinking, despite existing standards?

Table 2.10 summarizes the analytical assessment regarding organic and mineral substances, as well as emerging substances, which can cause health problems when wastewater is reused in applications where the consumer is in direct contact.

Analytical classification	Constituents
Conventional	SS, colloidal solids, BOD5, COD, TOC, ammonia, nitrates, nitrites, phosphorus Bacteria, viruses, protozoa
Non-conventional	Refractory organic substances, detergents, metals, dissolved salts
Emerging	Drug residues, pharmaceutical products, antibiotics for humans and for animals, hormonal estrogens, endocrine disruptors, industrial residues, microplastics

Table 2.10. *Predictive analytical assessment for treated wastewater reuse*

Concentrations of non-conventional and emerging constituents are in the range of micro- or nanogram per liter.

When considering the beneficial reuse of wastewater, the effect of these substances on public health must be taken into account.

Since it is not possible to monitor all non-conventional and emerging matter in the recovered water individually, some states have chosen TOC and total organic halogen (TOX) as alternative measurements for them.

Florida requires monitoring of TOC and TOX levels in recovered water when the recovered water is used for groundwater recharge or indirect reuse for potable application. Florida requires that the TOC and TOX concentrations in recovered water do not exceed 3.0 $mg \cdot L^{-1}$ and 0.2 $mg \cdot L^{-1}$, respectively, as a monthly average, and that a single sample must not exceed, respectively, 5.0 $mg \cdot L^{-1}$ and 0.3 $mg \cdot L^{-1}$.

2.3. References

Barbier, G. (2011). Perturbateurs endocriniens, le temps de la précaution. Parliamentary Report No. 765, July 12, Paris.

Birks, R., Colbourne, J., Hills, S., Hobson, R. (2004). Microbiological water quality in a large in-building water recycling facility. *Water Sci. Technol.*, 50(2), 165–172.

Bitton, G. (2005). *Wastewater Microbiology*. John Wiley & Sons, New York.

Blumenthal, U.J., Mara, D.D., Peasey, A., Ruiz-Palacios, G., Scott, R. (2000). Guidelines for the microbiological quality of treated wastewater used in agriculture: Recommendations for revising WHO guidelines. *Bull. World Health Organ.*, 78(9), 1104–1116.

Chauret, C., Springthorpe, S., Sattar, S. (1999). Fate of Cryptosporidium oocysts, Giardia cysts, and microbial indicators during wastewater treatment and anaerobic sludge digestion. *Can. J. Microbiol.*, 45, 257–262.

Chu, W.H., Wang, J.Y., Kao, C.M. (2003). A simplified risk-based approach for process screening in municipal wastewater reclamation and reuse. *Water Sci. Technol.*, 47(1), 257–262.

Cloete, T.E., Rose, J., Nel, L.H., Ford, T. (2004). *Microbial Waterborne Pathogens*. IWA Publishing, London.

Crook, J. (1998). Water reclamation and reuse criteria. In *Wastewater Reclamation, Recycling and Reuse*, Asano, T. (ed.). US EPA, New York.

Devaux, I., Gerbaud, L., Planchon, C., Bontoux, J., Glanddier, P.Y. (2001). Infectious risk associated with wastewater reuse: An epidemiological approach applied to the case of Clermont-Ferrand, France. *Water Sci. Technol.*, 43(12), 53–60.

Doran, G., Morris, D., O'Hare, C., DeLappe, N., Bradshaw, B., Corbett-Feeney, G., Cormican, M. (2005). Cost-effective application of pulsed-field gel electrophoresis to typing of Salmonella enterica Serovar Typhimurium. *Appl. Environ. Microbiol.*, 71(12), 8236–8240.

Emerson, K., Russo, R.C., Lund, R.E., Thurston, E.V. (1975). Aqueous ammonia equilibrium calculations: Effect of pH and temperature. *Journal of the Fisheries Research Board of Canada*, 32(12), 75–85.

Garcia, A., Yanko, G., Widmer, G. (2002). Giardia cysts in tertiary-treated wastewater effluents: Are they infective? *Water Environ. Res.*, 74, 541–544.

Gardner, E.A. (2003). Some examples of water recycling in Australian urban environments: A step towards environmental sustainability. *Water Sci. Technol. Water Supply*, 3(4), 21–31.

Gennaccaro, A., McLaughlin, M.R., Quintero-Betancourt, W., Huffman, D., Rose, J.B. (2003). Infectious *Cryptosporidium parvum* oocysts in final reclaimed effluent. *Appl. Environ. Microbiol.*, 69(8), 4983–4984.

Gerba, C.P. and Rose, J.B. (2003). International guidelines for water recycling: Microbiological considerations. *Water Sci. Technol. Water Supply*, 3(4), 311–316.

Haas, C.N. (2000). Epidemiology, microbiology, and risk assessment of waterborne pathogens including Cryptosporidium. *J. Food Prot.*, 63(6), 827–831.

Howard, G., Pedley, S., Tibatemwa, S. (2000). Quantitative microbial risk assessment to estimate health risks attributable to water supply: Can the technique be applied in developing countries with limited data? *J. Water Health*, 4(1), 49–65.

Huot, C., Gauvin, D., Levesque, B., Payment, P. (2008). *Portrait québécois et évaluation du risque à la santé de la réutilisation des eaux usées domestiques traitées pour l'irrigation de grandes surfaces gazonnées*. Institut national de la santé publique, Québec.

Kolpin, D.W., Furlong, E.T., Meyer, M.T., Thurman, E.M., Zaugg, S.D., Barber, L.B., Buxton, H.B. (2002). Pharmaceuticals, hormones, and other organic wastewater contaminants in U.S. streams, 1999–2000: A national reconnaissance. *Environ. Sci. Technol.*, 36(6), 1202–1211.

Krikelis, V., Spyrou, N., Makoulatos, P., Serie, C. (1984). Seasonal distribution of enteroviruses and adenoviruses in domestic sewage. *Can. J. Microbiol.*, 31, 24–25.

Law, I.B. (1996). Rouse Hill – Australia's first full scale domestic non-potable reuse application. *Water Sci. Technol.*, 33(10/11), 71–78.

Lazarova, V., Hills, S., Birks, R. (2003). Using recycled water for non-potable, urban uses: A review with particular reference to toilet flushing. *Water Sci. Technol. Water Supply*, 3(4), 69–77.

Lopez, B. and Laurent, B. (2013). Campagne exceptionnelle d'analyses des substances présentes dans les eaux souterraines. Report, ONEMA – BRGM, BRGM/RP 61853 FR, Paris.

Lopman, B.A., Reacher, M.H., van Duijnhoven, Y., Hanon, F.-X., Brown, D., Koopmans, M. (2003). Viral gastroenteritis outbreaks in Europe (1995–2000). *Emerg Infect Dis.*, 9, 90–96.

Marsalek, J., Schaefer, K., Exall, K., Brannen, L., Aidun, B. (2002). Water reuse and recycling. Report, Canadian Council of Ministers of the Environment, Winnipeg, Manitoba.

Maunula, L., Miettinen, I., von Bonsdorff, C.-H. (2005). Norovirus outbreaks from drinking water. *Emerg. Infect. Dis.*, 11(11), 1715–1721.

Michael-Kordatou, I., Michael, C., Duan, X., He, X., Dionysiou, D.D., Mills, M.A., Fatta-Kassinos, D. (2015). Dissolved effluent organic matter: Characteristics and potential implications in wastewater treatment and reuse applications. *Water Research*, 77, 213–248.

Olivieri, A.W., Eisenberg, D.M., Soller, J.A., Suffet, I.H., Froines, J., Cooper, R.C., Gagliardo, P.F. (eds). (1998). Comparative chemical risk assessment of repurified water from an advanced water treatment facility and a raw drinking water supply. In *Water Reuse '98 Conference Proceedings*. American Water Works Association, Denver.

Ottosson, J. (2002). Faecal contamination of greywater – Assessing the treatment required for hygienically safe reuse or discharge. In *2nd International Symposium on Ecological Sanitation*, Lübeck.

Ottosson, J. and Stenstrom, T.A. (2003). Faecal contamination of greywater and associated microbial risks. *Water Res.*, 37, 645–655.

Payment, P., Plante, R., Cejka, P. (2001). Removal of indicator bacteria, human enteric viruses, Giardia cysts, and Cryptosporidium oocysts at a large wastewater primary treatment facility. *Can. J. Microbiol.*, 47, 188–193.

Rose, J.B., Haas, C.N., Regli, S. (1991). Risk assessment and control of waterborne giardiasis. *Am. J. Public Health*, 81(6), 709–713.

Rose, J.B., Dickson, L.J., Farrah, S.R., Carnahan, R.P. (1996). Removal of pathogenic and indicator microorganisms by a full-scale water reclamation facility. *Water Res.*, 30(11), 2785–2797.

Rose, J.B., Farrah, S., Harwood, V.J., Levine, A.D., Lukasik, J., Menendez, J., Scott, T.M. (2004). Reduction of pathogens, indicator bacteria and alternative indicators by wastewater treatment and reclamation processes. Final Report 00-PUM-2T, Santé Canada Water Environment Research Foundation/International Water Association.

Santé Canada (2010). Recommandations canadiennes sur les eaux domestiques recyclées destinées à alimenter les chasses d'eau des toilettes et des urinoirs. Report, Government of Canada, Ottawa, Ontario.

Schaub, S.A. and Oshiro, R.K. (2000). Public health concerns about caliciviruses as water-borne contaminants. *J. Infect. Dis.*, 181(2), S374–S380.

Scott, T.T., McLaughlin, M.R., Harwood, V.J., Chicukula, V., Levine, A., Gennaccaro, A., Lucasik, J., Farrah, S.R., Rose, J.B. (2003). Reduction of pathogens, indicator bacteria and alternative indicators by wastewater treatment and reclamation processes. *Water Sci. Technol. Water Supply*, 3(4), 247–252.

Storey, M.V. and Ashbolt, N.J. (2003). A risk model for enteric virus accumulation and release from recycled water distribution pipe biofilms. *Water Supply*, 3(3), 93–100.

Toze, S. (2004). Literature review on the fate of viruses and other pathogens and health risks in non-potable reuse of stormwater and reclaimed water. Report, Australian Water Conservation and Reuse Research Program, Commonwealth Scientific and Industrial Research Organization.

Trevett, A.F., Carter, R.C., Westrell, T., Bergstedt, O., Stenstrom, T.A., Ashbolt, N.J. (2003). A theoretical approach to assess microbial risks due to failures in drinking water systems. *Int. J. Environ. Health Res.*, 13(2), 181–197.

Westrell, T., Schonning, C., Stenstrom, T.A., Ashbolt, N.J. (2004). QMRA (quantitative microbial risk assessment) and HACCP (hazard analysis and critical control points) for management of pathogens in wastewater and sewage sludge treatment and reuse. *Water Sci. Technol.*, 50(2), 23–30.

Yates, M. and Gerba, C. (1998). Microbial considerations in wastewater reclamation and reuse. In *Wastewater Reclamation and Reuse*, Asano, T. (ed.). Technomic Publishing Co., Lancaster/CRC Press, New York.

York, D.W. and Walker-Coleman, L. (2000). Pathogen standards for reclaimed water. *Water Environ. Technol.*, 12(1), 58–61.

3

Applications and Uses of Reused Water

The reuse of treated municipal wastewater is an interesting practice in terms of freshwater conservation, which is rare in many countries around the world, due to, among other things, climate change, pollution and past and current sub-optimal management practices. Projections for 2035 indicate a potential moderate to high water stress for about two-thirds of the world's population. Water scarcity, one of the world's major problems – and millions of people without access to fresh water – is likely to dramatically increase. The beneficial use of municipal and industrial wastewater, as well as the increasing demands on limited water resources, have led to the emergence of wastewater recovery and reuse as an integral part of water resources. This solution is the most developed in terms of volume in regions facing such problems, like the Middle East, Australia and the Southwestern United States. The reuse of wastewater is a competitive alternative in countries where wastewater management is an important issue for environmental or practical reasons such as the state of Florida, in tourist and coastal regions such as France, Spain and Italy, as well as in densely populated countries such as Germany and England. The reuse of treated wastewater is growing very fast, and volumes have increased from ~10% to 29% a year in Europe, the United States, China, and up to 41% in Australia (Aziz and Farissi 2014). China is the leading country in Asia for treated wastewater reuse with an estimated area of 1.3 Mha, including Vietnam, India and Pakistan (Zhang and Shen 2017). Currently, it is estimated that only 37.6% of the urban wastewater in India is treated (Singh et al. 2019). Using 90% of recovered water, Israel is the largest user of treated wastewater for the irrigation of agricultural land (Angelakis et al. 1999; Angelakis and Snyder 2015).

Water reuse can provide an alternative to existing systems and ensure water availability. It can be planned or unplanned. In the unplanned reuse of water, the water source consists largely of previously used water. For example, current

For a color version of all figures in this chapter, see www.iste.co.uk/gaid/reuse1.zip.

drinking water is most often taken from rivers and streams into which previously treated wastewater has been discharged.

Planned water reuse involves systems designed to make use of treated wastewater.

One of the main catalysts for the evolution of wastewater recovery and reuse has been the need to provide alternative water resources to meet water needs for irrigation, industry, urban applications of non-drinking and drinking water due to growth and development in many parts of the world (Table 3.1).

Applications and uses of treated wastewater	Constraints
Agriculture and landscape irrigation Crop irrigation Nurseries Golf courses Green belts Public gardens Residential areas Play areas	Surface and groundwater pollution, if not properly managed Crop marketing and public acceptance Effect of water quality, especially salts, on soils, grasses and crops Public health issues related to pathogens (bacteria, viruses and parasites of residential areas)
Urban use Floor washing Bus and car washing Street cleaning	Pathogens transmitted by aerosols
Reuse in the industry Cooling water Process water Boiler feed Power plants, refineries, mills, etc.	Wastewater components cause scaling, corrosion, and biological growth in boiler feed and fouling
Groundwater recharge	Organic chemicals in recovered wastewater and their toxicological effects Saltwater intrusion Total dissolved salts, nitrates, pathogens (bacteria and viruses) and micropollutants
Recreational and environmental areas Increased flow Feeding of artificial lakes and groundwater or coastal aquifers Renovation and rehabilitation of areas Fishing grounds Sports and leisure areas Snow making	Pathogens (bacteria and viruses) Eutrophication due to nitrogen and phosphorus in receiving waters Toxicity to aquatic life

Non-drinking water Fire hydrants Flush water from toilets Air conditioning	Pathogens transmitted by aerosols Effects of water quality on scaling, corrosion, biological growth and fouling
Drinking water Direct supply Mixing in tanks, drinking water tanks Food processing industries Ultrapure water	Micropollutants Industrial residues Toxicological effects Health concerns regarding the transmission of pathogens, in particular viruses Public acceptance
Worksites and construction Dust control or surface cleaning of roads, construction sites and other traffic areas Concrete mixing and other construction processes	Pathogens transmitted by aerosols

Table 3.1. *Different applications and uses of treated wastewater for its reuse*

Water reuse (also known as water recycling or water recovery) recovers water from various sources, and then treats it and reuses it for beneficial purposes such as agriculture and irrigation, drinking water supply, groundwater recharge, industrial processes and environmental restoration. Water reuse can provide alternatives to existing water supplies and be used to improve water safety, sustainability and resilience.

The uses of treated wastewater reuse define the degree of treatment required, technical solutions are implemented according to a multi-barrier approach, combining treatment and different environmental and health protections.

Water reuse can be defined as planned or unplanned. Unplanned water reuse refers to situations in which a water source is essentially made up of previously used water. The planned reuse of water refers to water supply systems designed to advantageously reuse a recycled water supply source. Often, communities will seek to optimize their overall water use by reusing water as much as possible within the community before water is reintroduced into the environment. Examples of planned reuse include agricultural and landscape irrigation, industrial process water, the supply of drinking water and the management of groundwater supplies.

The serious drought issues that have been affecting Spain for several years have highlighted the importance of agricultural reuse for the sustainable development of Catalonia, Andalusia, and the Canary and Balearic Islands.

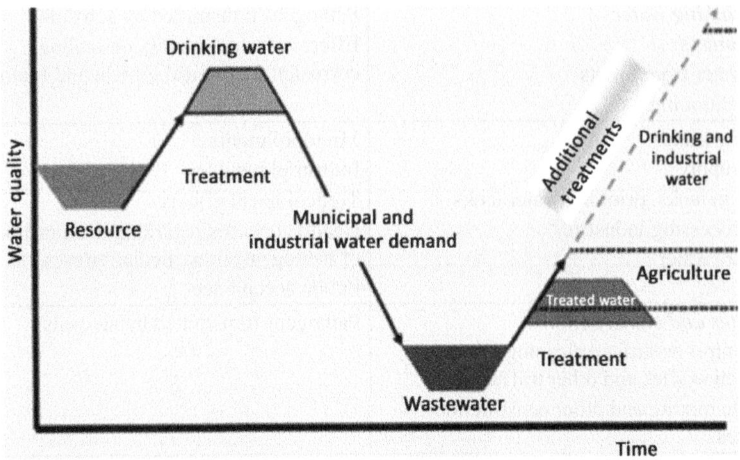

Figure 3.1. *General pattern of water demand to satisfy water deficits*

Water sources likely to be reused include wastewater, industrial treatment and cooling water, rainwater, agricultural runoff water and water produced by natural resource extraction activities. These water sources are treated in an appropriate manner to meet the characteristics appropriate for a particular use (Figure 3.1). The characteristics appropriate for the use are the treatment requirements for bringing water to the required quality to ensure public health, environmental protection or the specific needs of users.

For example, water recovered for crop irrigation should be of sufficient quality to prevent damage to plants and soil, maintain food safety and protect the health of agricultural workers. In uses where human exposure is higher, water may require further treatment.

Depending on the quality requirements of users (Figure 3.2), two main reuse classes can be defined:

– non-drinking uses, mainly in the agricultural, industrial and urban sectors;

– drinking uses, which after a thorough treatment can be "direct" or "indirect" following a passage in the natural environment.

Non-drinking uses are mainly:

– agricultural irrigation: cereal crops, fruit trees, forests, frost protection, etc.;

– industrial uses: cooling water, process water, steam generators, equipment cleaning, fire protection, etc.;

– urban uses: maintenance of roads, toilet flushing, buses and cars, cleaning of public ornamental buildings, irrigation of parks, cemeteries, public gardens, etc.;

– recreational uses: lakes and artificial basins, support for low water flow in rivers, maintenance of natural habitats and wetlands, snow production, fisheries, etc.;

– groundwater recharge: fight against the intrusion of seawater or brackish water in the case of overexploitation of the aquifer, replenishment of critical groundwater, storage of treated water for future uses.

Drinking uses are mainly:

– the indirect production of drinking water: increase in the availability of water, a mode of production often linked to the groundwater recharge;

– direct production of drinking water.

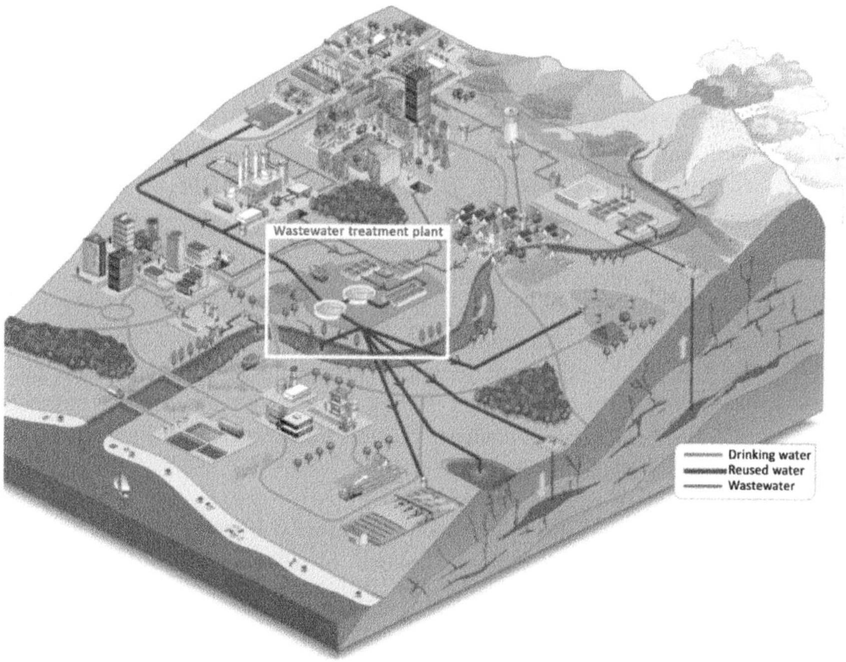

Figure 3.2. *General pattern of reused water uses*

3.1. The different uses

3.1.1. *Agriculture*

The reuse of wastewater in agriculture is a practice that is thousands of years old. As such, the agricultural sector is well known to be the largest user of water, accounting for almost 70% of global water consumption. The fact that around 20 million ha in the world are irrigated with wastewater water suggests a major source of irrigation (Ecosse 2001).

In many arid and semi-arid countries, water is becoming an increasingly scarce resource and planners are forced to consider all sources of water that could be economically and efficiently used to promote a subsequent development. This problem will also be observed in many regions of the world, with climate change.

At the same time, with the population growing at a high rate, the need to increase food production is evident. Many countries have included treated wastewater reuse as an important dimension of water resource planning. In the driest regions of Australia and the US, wastewater is used for agriculture, freeing high-quality water reserves for drinking. Some countries, for example, the Hashemite kingdom of Jordan and the kingdom of Saudi Arabia, have a national policy for the reuse of all treated wastewater effluents and have already made significant progress in this regard. In China, the use of wastewater in agriculture has grown rapidly since 1958, and today more than 1.33 million ha are irrigated with wastewater effluents. It is generally accepted that the use of wastewater in agriculture is justified for agronomic and economic reasons.

Water deficit will mean that an alternative resource must be considered to compensate for the shortages and deficiencies. The quality of this resource must meet physicochemical and bacteriological conditions and must have all the characteristics that would not result in issues for health or equipment and farm operation. For example, brackish water cannot be used for agricultural purposes because of its high dissolved salt content. Treated wastewater must be checked for bacteria due to the associated health risks.

The availability of this resource close to urban areas will increase crop choice for farmers. The nitrogen and phosphorus content of wastewater could reduce or eliminate the need for commercial fertilizers.

However, most of the wastewater used for irrigation is not treated (Scott et al. 2004, 2009; Jiménez and Asano 2008). In some countries, partially treated or untreated wastewater is used for irrigation (Scott et al. 2004). Untreated wastewater is widely used for agriculture in many countries. This is one of the world's major environmental and public health concerns. Instead of using untreated wastewater, treated wastewater has proved to be a more applicable and environmentally friendly option. In addition, environmental toxicity due to exposure to solid waste is also one of the main health issues. Untreated wastewater often contains a wide range of chemical contaminants from landfills, chemical wastes from industrial discharges, heavy metals, fertilizers, textiles, leather, paper, food processing waste and pesticides. The World Health Organization (WHO) has warned of significant health consequences from direct use of wastewater for irrigation (WHO 1989, 2006; Carr 2005; ANSES 2012).

Currently, the use of antibiotics has exceeded the scope of their applications to humans. Their application in livestock farming, biochemical industries and agriculture has become a common practice. The intensive use and/or misuse of antibiotics have given rise to multidrug-resistant bacteria, which carry multiple resistance genes (Monchalin and Aviron-Violet 2000; ANSES 2012; Lyu et al. 2016; Tripathi and Tripathi 2017). These multi-resistant bacteria are discharged into the sewage system and collected in wastewater treatment plants. The latter serve as a point of accumulation and development for antibiotic-resistant bacteria (ARB) and antibiotic resistance genes (ARG). These multidrug-resistant bacteria can be spread to different bacterial species via mobile genetic elements and horizontal gene transfer (Goldstein 2013; Gupta and Diwan 2017). Previous studies have shown that some pathogens can survive in wastewater even during and after treatment processes, in particular the methicillin-resistant *Staphylococcus aureus* (MRSA) and vancomycin-resistant enterococci (VRE) (Caplin et al. 2008; Börjesson et al. 2009). Use of treated wastewater in irrigation provides favorable conditions for the growth and persistence of total coliforms and fecal coliforms (Akponikpe et al. 2011; Sacks and Bernstein 2011). In addition, studies have reported the presence of various bacterial pathogens, such as *Clostridium, Salmonella*, streptococci, viruses, protozoa and helminths in crops irrigated with treated wastewater (Carey et al. 2004; Mañas et al. 2009; Samie et al. 2009). Survival of ARB in treated wastewater can lead to serious health risks for people exposed to reused water (Goldstein 2013).

These contaminants pose health risks to communities (farmers, farm workers, their families and consumers of wastewater-irrigated crops) living near wastewater sources and irrigated areas with untreated wastewater (Qadir et al. 2016). Wastewater also contains a wide variety of organic compounds. Some of them are toxic or carcinogenic and have adverse effects on the embryo (Jarup 2003; Shakir et al. 2016).

Reuse in agriculture has the advantage of having nutrients present in wastewater and thus reducing the demand for commercial fertilizers. Water requirements for irrigation vary with the seasons; a storage system or combination with other reuse schemes are often planned. Agricultural irrigation is the largest consumer of recycled water with many well-recognized benefits and advantages, including contribution to food security. In arid and semi-arid countries, the reuse of wastewater provides most of the irrigation water.

In some arid and semi-arid countries, reused water provides most of the irrigation water. In addition to chronic water shortage, the need for this resource has accelerated in recent years due to frequent droughts in the arid areas of the Sahel, the United States, the Mediterranean countries, the Middle East and South Asia. At the moment, it is therefore a question of alleviating water deficits and ensuring agricultural production by providing sufficient water for irrigation. This solution has thus become a priority for many countries and regions. While this discussion focuses on the planned reuse of wastewater for agriculture, the reality is that in many parts of Asia, Central America and Africa, wastewater is used in an unplanned manner to grow crops. These crops, which are at the heart of farmers' livelihoods and have considerable importance for local communities, could not be grown without this source of water and nutrients. It is not realistic to expect that this wastewater be treated; the question is therefore how to continue applying the effluent with minimum risk to human health and soil productivity.

3.1.1.1. *Irrigation methods*

The most common use of wastewater is by far irrigation. Irrigation of crops or green spaces is the most common way to reuse urban wastewater. At a global level, it is also the solution which is the most promising in the short and medium term. Many different methods are used by farmers to irrigate crops. This ranges from watering individual plants with a water can to highly automated irrigation with a central pivot system. However, from the ground wetting point, these methods can be grouped under five categories.

Flood irrigation: water is applied to the entire field to seep into the ground (for example, wild flooding, contour flooding, curbs, ponds, etc.).

If the recovered water is to be used for irrigating forage crops, large crops (cereals, industrial crops) or forest trees, the quality of the secondary effluent may be sufficient subject to disinfection, however. For vegetables intended to be consumed raw, additional wastewater treatment steps are essential for consumer protection.

Crop irrigation is the first objective of treated wastewater reuse. As a result, the health risks and soil damage can be high if wastewater is only partially treated (simple settling) or misused (non-compliant discharges).

Furrow irrigation: water is applied between the ridges (e.g. leveled furrows, contour furrows and undulations). The water reaches the ridge, where the roots of plants are concentrated, by capillarity.

Sprinkler irrigation: water is applied in the form of spray and reaches the ground a bit like rain (e.g. portable and stationary sprinklers, mobile sprinklers, spray guns and center-pivot systems). The application rate is adjusted so that no puddles are created on the surface.

It is the technique most concerning to the health of consumers if the crops are consumed raw. The risk of contamination can be very high if any of the water treatments such as disinfection is not well controlled.

While deep irrigation can harm groundwater quality, sprinkling creates aerosols that can be a nuisance to users. Consumers of raw vegetables and agricultural workers are exposed to contamination by undisinfected wastewater. Helminths (*Ascaris*, trichocephalus, ankylostoma, etc.) represent the main microbiological risk, followed by bacterial and viral diseases.

Figure 3.3. *Examples of sprinkling of an environmental area and a field*

Sprinkler irrigation should be avoided if the WHO guidelines are not fully met. Even with treated wastewater, sprinkler irrigation can promote the transmission of virus aerosols, although this is probably rare.

Root irrigation: water is applied under the root zone in such a way that the root zone is wetted by capillary rise (e.g. underground irrigation channels and buried pipes). Deep surface channels or buried pipes are used for this purpose.

Localized irrigation: water is applied around each plant or group of plants in such a way that it locally wets only the root zone (e.g. drip irrigation, bubblers and micro-sprinklers). The application rate is adjusted to meet evapotranspiration needs to minimize percolation losses. For the protection of agricultural workers, drip irrigation is recommended where possible as a replacement for sprinkling and furrow irrigation. Drip irrigation is the most appropriate method to prevent salinity problems, especially since this technique has the lowest contamination potential.

For all irrigation techniques, depending on the quality of the treated wastewater, use may be restricted or unrestricted:

– *restricted use*:

- water of poor quality in limited areas and for specific crops,

- restrictions are imposed based on the type of soil, proximity of the irrigated area to a drinking water aquifer. Irrigation methods, harvesting techniques and fertilizer application rates are monitored,

- generally applicable only to small quantities of wastewater, and applied in specific locations where areas and crops are well defined and not susceptible to change,

- farmer must be trained to manage poor-quality effluents;

– *unrestricted use:*

- use of high quality effluents, as a substitute for conventional water, to irrigate any crop on any type of soil,

- the contact does not pose a health risk,

- crops without any restrictions also include vegetables consumed raw.

The FAO (Food and Agriculture Organization of the United Nations) proposes a classification of crops according to the level of risk for consumers and farmers:

1) low risk:

- industrial crops not intended for human consumption (e.g. cotton and sisal); crops treated by heat or drying before being used for human consumption (e.g. cereals, nuts and sugar beets),

- vegetables and fruits grown exclusively for canning or another treatment that effectively destroys pathogenic germs,

- forage crops and other feed crops for animals that are harvested and sundried;

2) average risk:

- pastures and herbaceous forage crops,

- crops for human consumption which are not in direct contact with wastewater, provided that nothing is picked up on the ground and that sprinkler irrigation is not used (e.g. arboriculture and vineyards),

- crops for human consumption after cooking (e.g. potatoes, eggplant and cooked beet),

- crops for human consumption, whose skin is not eaten (e.g. citrus fruits and bananas),

- all crops not identified as "high risk" if sprinkler irrigation is used;

3) high risk:

- all food consumed raw or grown in close contact with effluents (e.g. lettuce and carrots),

- sprinkler irrigation, regardless of crop type, within 100 m of residential areas or public access venues.

The production of millions of farmers around the world depends on the availability of water. They use, depending on the case, drinking water, surface waters (rivers, streams, reservoirs, dams, etc.), groundwater, but also treated wastewater if available.

Global estimates of the extent of wastewater irrigation range from 4 to 20 million hectares.

Economic and environmental advantages	Risks for environment and public health if treatment is unsuitable, insufficient, or poorly used
– Reduces freshwater demand. – Provides a reliable water supply to farmers. – Reduces pollution of rivers, canals and other surface water. – Recycles organic matter and nutrients in soils, thus reducing the need for chemical fertilizers. – Increases crop yield and farmers' income.	– Health risks for irrigators. – Health risks for consumers of vegetables poorly irrigated with wastewater. – Contamination of groundwater (nitrates, micropollutants, pathogens, etc.). – Accumulation of chemical pollutants in the ground (salts, heavy metals, etc.). – Reduces water levels of rivers and streams.

Table 3.2. *Compromise between the advantages and risks of using wastewater in agriculture*

Figure 3.4. *Diversity of irrigation applications with treated wastewater*

3.1.1.2. *important quality parameter for wastewater agricultural use*

3.1.1.2.1. Irrigation constraints

Many schemes for the classification of irrigation water have been proposed. Irrigation water is classified into three groups based on salinity, sodium content, toxicity and various other parameters.

These general guidelines for water quality classification help identify potential agricultural production problems associated with conventional water use (Ayers and Westcot 1985, 1994). The main parameters that can affect plant growth are:

– the specific toxicity of metals in water;

– the osmotic pressure caused by high salt content in soil water;

– the dispersion of soil particles caused by high salt content in soil water;

– nutrients;

– in regions of the world where helminthic diseases caused by *Ascaris* and *Trichuris* spp. are endemic in the population and where raw untreated wastewater is used to irrigate salad crops and/or vegetables consumed raw, the transmission of these infections is significant through consumption of these crops (Shuval et al. 1985, 1986, 1999).

3.1.1.2.2. Salinity

The main criterion for assessing the quality of natural water for an irrigation project is its total concentration of soluble salts.

Salinity is the most important parameter in irrigation and electrical conductivity (EC) is its unit. The EC of water is used as an alternative measurement for the total dissolved salt (TDS) concentration in water and is expressed in $dS \cdot m^{-1}$ or $mmho \cdot cm^{-1}$. One $dS \cdot m^{-1}$ is equivalent to 1 $mmho \cdot m^{-1}$. In other cases, EC can be expressed in $\mu S \cdot cm^{-1}$.

$$\mu S \cdot cm^{-1} = 10^{-4}\ S \cdot m^{-1} \text{ and } dS \cdot m^{-1} = 1{,}000\ \mu S \cdot cm^{-1}$$

Salinity can also be expressed in mg of $TDS \cdot L^{-1}$.

The empirical relationship between EC and TDS concentration is:

$$\text{TDS } (mg \cdot L^{-1}) = \text{EC } (dS \cdot m^{-1} \text{ or } mmho \cdot cm^{-1}) \times 640$$

The effect of salinity on agricultural production must be considered first, when the recovered water is used for irrigation. When a high concentration of specific ions, and not the osmotic pressure alone, affects the growth of the crop, it is called “ion-specific toxicity”.

The consequences of excessive salinity in irrigation water are presented by two criteria:

– damage to the soil (indirectly on crops);

– crop damage.

Sodium is subject to significant accumulation in the soil, and therefore must be given special attention. Plants draw water from the soil, leaving a large part of the salts supplied by the watering water and thus increasing the salinity of the soil water. As the osmotic pressure of the soil water increases with its concentration in dissolved salts, the plant then dedicates most of its energy not to develop, but to adjust the salt concentration of its plant tissue so that it can extract from the soil the water it needs. An irrigation water quality scale according to their salinity and EC is proposed. The intrusion of seawater or brackish water into an estuary or groundwater that sometimes joins sanitation systems as parasite water results in an increase of the environment conductivity (Richards 1954).

Water quality	Water conductivity ($\mu S \cdot cm^{-1}$)	Soluble salts estimated in NaCl ($mg \cdot L^{-1}$)
I. Excellent	250	<160
II. Low salinity	250–750	160–500
III. Moderate to high salinity	750–2,250	500–1,500
IV. Very high salinity	2,250–5,000	1,500–3,600

Table 3.3. *Influence of salinity on irrigation water quality*

Parameters	Unit
Physical	
Total dissolved solids	$mg \cdot L^{-1}$
Electrical conductivity	$dS \cdot m^{-1}$
Temperature	°C
Turbidity	NTU
Color	°Hazen or $mg \cdot L^{-1}$ Pt/Co
Hardness	mg $CaCO_3 \cdot L^{-1}$
Suspended solids	$mg \cdot L^{-1}$
Chemical	
pH	–
Total dissolved salts	$mg \cdot L^{-1}$
Calcium	$mg \cdot L^{-1}$
Magnesium	$mg \cdot L^{-1}$
Sodium	$mg \cdot L^{-1}$
Carbonate	$mg \cdot L^{-1}$
Bicarbonate	$mg \cdot L^{-1}$
Chloride	$mg \cdot L^{-1}$
Sulfate	$mg \cdot L^{-1}$
Sodium absorption total	$mg \cdot kg^{-1}$
Boron	$mg \cdot L^{-1}$
Heavy metals per element	$mg \cdot L^{-1}$
Azote (ammonia, nitrate)	$mg \cdot L^{-1}$
Phosphate	$mg \cdot L^{-1}$
Potassium	$mg \cdot L^{-1}$

Table 3.4. *Parameters used to define water quality for agriculture*

As a guide, seawater has a conductivity of about 55,000 $\mu S \cdot cm^{-1}$, which is equivalent to a sodium chloride (NaCl) concentration of about 35 $g \cdot L^{-1}$, namely, water about 10 times saltier than class IV water. This value therefore characterizes the major disadvantage of using seawater for irrigation. As to wastewater, conductivity is much lower, as it is close to 1,100 $\mu S \cdot cm^{-1}$.

Some crops may be sensitive to a change in salinity, and it is important to take this parameter into account when treated wastewater is used for irrigation (Boutin et al. 2009). Another aspect of agricultural concerns is the effect of dissolved salts (TDS) in irrigation water on plant growth. Dissolved salts increase the osmotic potential of soil water and an increase in the osmotic pressure of the soil solution increases the amount of energy that plants must expend to absorb water from the soil. As a result, respiration is increased, and the growth and yield of most plants gradually decrease as osmotic pressure increases. Although most plants react to salinity according to the total osmotic potential of soil water, some plants are sensitive to specific ionic toxicity. The total salt concentration (for all practical purposes, total dissolved solids) is one of the most important parameters of agricultural water quality. Indeed, the salinity of soil water is linked to the salinity of irrigation water and often determined by it. As a result, plant growth, crop yield and product quality are affected by the total salts dissolved in irrigation water. Similarly, the rate of salt accumulation in soil, or soil salinization, is also directly affected by the salinity of irrigation water. Total concentration in salt is expressed in milligrams per liter ($mg \cdot L^{-1}$) or parts per million (ppm).

Many ions that are harmless or even beneficial at relatively low concentrations can become toxic to plants at high concentrations, either by direct interference with metabolic processes or by indirect effects on other nutrients, which could be made inaccessible.

3.1.1.2.3. Electrical conductivity

EC is widely used to indicate the total ionized constituents of water. It is directly linked to the sum of the cations (or anions), as chemically determined, and is closely correlated, in general, with the total salt concentration. EC is a quick and reasonably accurate determination, and values are always expressed at a standard temperature of 25°C to allow for the comparison of readings taken under varying climatic conditions. It should be noted that the EC of the solutions increases by about 2% per degree of temperature increase. The unit of EC is the $dS \cdot m^{-1}$.

3.1.1.2.4. Sodium adsorption rate (SAR)

Sodium is a unique cation due to its effect on soil. When present in the soil in an exchangeable form, it causes harmful physicochemical changes to the soil, especially to the soil structure. It has the ability to disperse soil when present above a certain threshold value, in relation to the TDS concentration. Soil dispersion results in reduced rates of water and air infiltration into the soil. Once dried, the dispersed soil forms crusts that are difficult to plough and interfere with germination and emergence of seedlings. Irradiation water could be a source of excess sodium in the soil solution and should therefore be assessed for this hazard. The most reliable indicator of sodium risk in irradiation water is the SAR. The SAR is defined by the formula:

$$\mathrm{SAR} = \frac{\mathrm{Na}}{\sqrt{\frac{\mathrm{Ca+Mg}}{2}}}$$

where the ion concentrations are expressed in $\mathrm{meq \cdot L^{-1}}$.

The effect of sodium ions in irrigation water on the reduction of soil infiltration rate and permeability depends on the concentration of sodium ions relative to the concentration of calcium and magnesium ions (as indicated by the SAR) and total salt concentration.

3.1.1.2.5. pH

pH is an indicator of water acidity or basicity, but it is rarely a problem in itself. The normal pH range for irrigation water is 6.5 to 8.4.

3.1.1.2.6. Trace elements and heavy metals

Some elements are normally present at relatively low concentrations, generally below a few mg/L, in conventional water and are called trace elements. They are not normally included in routine analysis of ordinary irrigation water, but should be considered when using wastewater effluents, particularly if contamination from industrial wastewater discharges is suspected. These include aluminum (Al), beryllium (Be), cobalt (Co), fluorine (F), iron (Fe), lithium (Li), manganese (Mn), molybdenum (Mo), selenium (Se), tin (Sn), titanium (Ti), tungsten (W) and vanadium (V). Heavy metals are a special group of trace elements that have been shown to create certain health risks when absorbed by plants. This group includes arsenic (As), cadmium (Cd), chromium (Cr), copper (Cu), lead (Pb), mercury (Hg) and zinc (Zn).

Element	Max. conc. ($mg \cdot L^{-1}$)	Toxicity
Al **Aluminum**	5.0	May cause non-productivity in acidic soils (pH < 5.5), but more alkaline soils with a pH > 7.0 will precipitate the ion and eliminate any toxicity.
As **Arsenic**	0.10	Plant toxicity varies widely, ranging from 12 $mg \cdot L^{-1}$ for Sudanese grass to less than 0.05 $mg \cdot L^{-1}$ for rice.
Be **Beryllium**	0.10	Plant toxicity varies widely, ranging from 5 $mg \cdot L^{-1}$ for kale to 0.5 $mg \cdot L^{-1}$ for dwarf beans.
Cd **Cadmium**	0.01	Plant toxicity varies widely, ranging from 5 $mg \cdot L^{-1}$ for kale to 0.5 $mg \cdot L^{-1}$ for dwarf beans.
Cr (VI) **Chrome**	0,10	Generally, not recognized as an essential growth element. Conservative limits recommended due to lack of knowledge about its toxicity to plants.
Cu **Copper**	0.20	Toxic to a number of plants at 0.1–1.0 $mg \cdot L^{-1}$ in nutrient solutions.
F **Fluoride**	1.0	Inactivated by neutral and alkaline soils.
Fe **Iron**	5.0	Non-toxic to plants in aerated soils but may contribute to soil acidification and loss of availability of essential phosphorus and molybdenum. Aerial watering can cause unsightly deposits on plants, equipment and buildings.
B **Boron**	1.0	Damages the leaves of fruit trees.
Li **Lithium**	2.5	Tolerated by most crops up to 5 $mg \cdot L^{-1}$; mobile in soil. Toxic for citrus at low concentration (<0.075 $mg \cdot L^{-1}$). Acts in the same way as boron.
Mn **Manganese**	0.20	Toxic to a number of crops at a few tenths to a few $mg \cdot L^{-1}$, but usually only in acidic soils.
Mo **Molybdenum**	0.20	Non-toxic to plants at normal concentrations in soil and water. Toxic to livestock if forage is grown in soils with high concentrations of available molybdenum.
Ni **Nickel**	0.20	Toxic to a number of plants from 0.2 $mg \cdot L^{-1}$.
Pb **Lead**	5.0	Can inhibit plant cell growth at very high concentrations.
V **Vanadium**	0.50	Toxic to many plants at relatively low concentrations.
Zn **Zinc**	2.0	Toxic to many plants at highly variable concentrations; reduced toxicity at pH > 6.0 and in fine-textured or organic soils.

Table 3.5. *Maximum permissible concentrations for crop irrigation (Ayers and Westcot 1985, 1994)*

Among them, some elements are needed for the development of crops such as boron, iron, zinc, copper and molybdenum. Wastewater most often contains these elements in trace amounts. They are retained in the surface layers of the soil by adsorption or ion exchange. An excess of metals can have harmful consequences on crops through degradation of leaves and/or roots (Faby 1997; Faby et al. 1997). Indeed, irrigation water containing certain ions at concentrations above threshold values can cause toxicity problems for plants. Toxicity is usually reflected in impaired growth, reduced yield, changes in plant morphology and even death of the plant. The degree of damage depends on the crop, its stage of growth, concentration of the toxic ion, climate and soil conditions.

Heavy metals pose the greatest long-term environmental risk, as they accumulate in the upper soil. If a soil is very permeable or the groundwater is close to the surface, contamination may occur. Many heavy metals and pollutants such as chlorinated hydrocarbons can accumulate in sufficient concentrations in crops and pose a serious threat to the health of livestock and humans. In most cases, harmful levels of chemicals can be avoided by prohibiting the use of untreated wastewater. The health risk depends on the method of application of irrigation water. Flooding involves the least investment, but the greatest risk for field workers if wastewater is not fully treated.

Municipal wastewater effluents can contain a number of toxic elements, including heavy metals due to industrial spills. Beyond certain concentrations, they can affect plant, foliage and root growth. Therefore, municipal wastewater effluents must be tested for trace element toxicity, especially when trace element contamination is suspected. Table 3.5 shows the maximum permissible concentrations in irrigation water and the toxicity caused by an excess of metal concentration.

3.1.1.2.7. Boron case

Boron is an element naturally very present in the environment, and therefore deserves special attention. It is found in combination with oxygen or other elements in compounds called borates. Boron is found in wastewater, mainly from detergents, industrial discharges or desalinated seawater. Borates are adsorbed on soil particles. The degree of adsorption depends on the type of soil. Boron is a constituent of important compounds in the plant. It performs functions similar to those of phosphorus. It is essential in areas of active plant growth, such as root tips, and in the development of new leaves and buds. It enables cell differentiation and contributes to the transport of substances. Boron compounds regulate plant-specific growth substances. A boron deficiency damages the plant tissue and thus inhibits growth. It is often the growing tip that is affected. Flowers and leaves curl up. Chlorosis appears on young leaves, their growth is distorted and twisted, and they turn brown. Shoots brown and die, flowers are deformed. Roots are short and brown, and their growth limited. However, when its concentration exceeds 1.0 mg/L, it can be toxic to the most sensitive plants. If there is too much boron, leaves curl up as in a potassium deficiency and the leaves at the bottom of the plant begin to turn yellow and fall off. The young leaves are chlorotic. The tips of the leaves turn yellow, then necrotize and fall off. This can affect leaves, twigs, inflorescences, rootlets: poor tissue development, discoloration, disturbed growth. Symptoms are observed on leaves and are characterized by discolored areas on the edge of the leaf blade. In serious cases, leaves remain small and stunted.

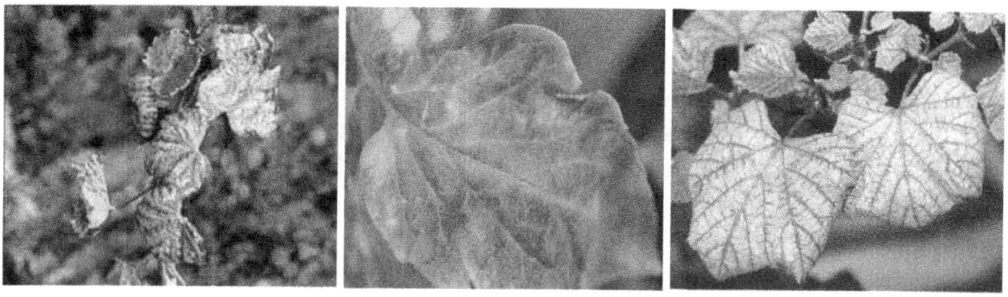

Figure 3.5. *Excess boron influence on plants*

3.1.1.2.8. Nutrients

Waste water from treatment plants contains many nutrients, especially nitrogen and phosphorus, which are beneficial for crops. However, excessive nutrient inputs can cause many problems. Excessive nitrogen application during the last part of the crop growing period can actually lead to overgrowth, delayed or uneven ripening, or a reduction in crop quality. If alternative water with a low nitrogen content is available, a change of water supply or a mixture of reclaimed water with other water supplies is recommended to control the nitrogen supply. Where this is not possible, nitrogen and phosphorus treatment must be implemented at the wastewater treatment plant by changing operating conditions.

3.1.2. *Reuse in industry*

The reuse of wastewater is increasingly being practiced to supplement the water requirements of industry.

Typically, the demand for wastewater for industry includes three cases:

– direct reuse with little or no treatment;

– the reuse of wastewater after an appropriate treatment;

– the reuse of wastewater involving additional treatment until drinking water standards are achieved.

The reuse of wastewater for industry has the double advantage of a generally continuous demand and the ability to use wastewater directly from wastewater treatment plants.

The main factors influencing industrial treated wastewater reuse include availability and reliability of this resource, and the water quality required by industrial processes.

The main uses in industry are for the increase in cooling system, process water, boiler feed water, wash water and various applications such as fire protection and dust control. Industry is the second largest user, accounting for about 25% of global demand. The sectors that use most water are thermal and nuclear power plants (cooling water) and paper mills.

The reuse of domestic wastewater and the internal recycling of industrial wastewater are well-established practices. Although there is a wide range of types of water reuse in industry, the main uses are as follows:

– closed or open circuit cooling systems;

– wash water;

– boiler feed;

– process water;

– various other uses such as fire protection and cleaning.

The requirements and application areas of water recycling in industry differ according to the type of industry, specific industrial processes and their performance targets.

The quality of reused water is well regulated and depends on the type of application or industrial production.

The quality requirements for cooling towers are imposed by the accumulation of salts and other impurities in these systems which operate in a closed loop, caused by water evaporation.

Indeed, on the surfaces of heat exchangers, scale deposits reduce the capacity of heat transfer. At high concentrations, they tend to accelerate corrosion and activate clogging.

Legionella bacteria are an increasingly crucial problem in cooling systems. They are stimulated, for example, by the presence of nutrients in wastewater, which accelerates their growth. Other microorganisms may grow in these systems due to the presence of nutrients and high temperatures (Monchalin and Aviron-Violet 2000; Carre 2017).

The most used method to limit bacterial growth in cooling systems is the addition of biocides. Once-through cooling involves a large amount of water but does not require a quality as rigorous. The quality of the discharged water is not changed, except for the temperature, which increases at the outlet.

Boiler feeding requires a reduction in water hardness and alkalinity to avoid the formation of scale and deposits in appliances. Other minerals can cause scaling such as sulfates, phosphates and silica. The most common types of scaling are those of calcium phosphate and calcium sulfate.

Requirements regarding the quality of the treatment water depend heavily on the type of industry and its manufacturing processes. Generally, it is desirable to minimize the concentration of suspended solids, turbidity and color. For other industries, it will be silica, iron, aluminum, hardness, microorganisms and pH. Finally, for some of them, and particularly the food industry, water quality must meet drinking water standards.

3.1.3. *Groundwater recharge*

The most common practices for indirect reuse of treated wastewater for drinking water production include:

– indirect (via infiltration ponds) or direct (via injection wells) groundwater recharge;

– refilling of surface tanks used for drinking water production (Degrémont 2005).

Groundwater recharge is used to preserve groundwater levels and protect coastal aquifers from saltwater intrusion to store treated wastewater. The recharge methods commonly used are surface spreading in reservoirs, direct injection into underground aquifers and infiltration after bank impermeability. Indeed, underground aquifers can provide a means of storage and availability for reuse. When the recovered water is used for irrigation, which has a high seasonal demand, this water can be stored in the underground aquifer during the period of low demand, which eliminates the need for large surface storage facilities.

The use of underground aquifers for storage purposes solves problems associated with open ponds such as odor production, evaporation losses, algal blooms, causing deterioration in water quality.

Wastewater at the outlet of treatment plants must undergo an additional treatment to eliminate SS, micropollutants and organic matter, as the risk of injecting wastewater without further treatment is to contaminate groundwater, which is often of very good quality. As such, a prudent approach should be adopted when an aquifer recharge is carried out with a direct supply of wastewater at the outlet of a treatment plant.

Due to the high health risks associated with possible microbiological or chemical pollution, treatment processes are based on the concept of "multiple barriers" to achieve and ensure a very high degree of efficiency and reliability. The very strict safety requirements for public health are ensured not only by adequate treatment, but also through regular and continuous monitoring of water quality and reliability of each individual process and the application of good project management practices (Degrémont 2005).

Particular attention should be paid to contamination by pathogenic microorganisms.

Frequent monitoring of groundwater quality is essential, as it takes up to 100 years for soil and groundwater to recover once they have been contaminated. Even low concentrations of micropollutants in treated wastewater can be an obstacle to the recharge of aquifers. In small amounts, these elements pose risks of short-term human toxicity and longer term diseases.

The concept of groundwater recharge includes storage, recovery and pumping of water for reuse.

Florida imposes limits for TOC and TOX (halogenated organic compounds) in reclaimed water for groundwater recharge to reduce the risk of organic and emerging substances.

The reused water can be introduced into underground areas through stabilization basins. This method is the simplest, oldest and most widely used for groundwater recharge. Water flows from the spreading basin through an unsaturated groundwater area. Percolation through the soil removes suspended solids.

The infiltration pond is another solution quite often implemented. The ponds are built on very permeable soils in order to accelerate percolation. They allow for a more efficient use of space and require little maintenance. The treated wastewater, filtered naturally through the soil, descends into the groundwater. Soil materials are thin enough for biochemical and microbiological reactions to occur on the materials along the water course.

3.1.4. *Recreational and environmental areas*

There are three main uses for treated wastewater reuse for recreational purposes, each requiring specific qualities:

– direct contact for swimming and water sports areas;

– snow generation;

– possible or accidental contact for boating and fishing;

– no contact like ornamental fountains (no public access), aquaculture, ponds, dams, lakes and water bodies.

The water quality criteria for direct contact are the most stringent. Water must be aesthetically pleasing and have acceptable physical and chemical properties for parameters such as color, taste and smell. Most importantly, the water must be free of toxic substances and microbiologically safe.

Water quality requirements are less strict for accidental contact.

Those relating to no-contact areas require a reasonable temperature to maintain aquatic life, sufficient dissolved oxygen concentration and acceptable chemical quality with a strong reduction of nutrients to avoid eutrophication and reasonable microbiological quality.

3.1.4.1. *Snow generation*

Reused water can be used to make artificial snow and the water quality must be high due to possible human contact. Snow Valley ski resort, located in the San Bernardino Mountains, makes snow from water recovered from the town of Running Spring, located about 15 km from the ski resort. The recovered water is transported by a pipeline and then stored in a reservoir where an aerator enriches the water with oxygen. The water is then used for snow-making on the ski slopes in winter and for irrigation to maintain a grass cover on the ski slopes and prevent erosion in summer.

In all direct contact uses, the protection of users is the most important requirement due to potential risks. As a result, water disinfection requirements are among the most stringent for non-drinkable uses and are similar to those of unrestricted irrigation in parks and green spaces open to the public. This level of treatment involves in all cases a filtration step followed by disinfection. Regular monitoring of water quality and control of distribution systems are recommended.

3.1.5. *Urban non-food uses*

In the context of a lasting drought, a solution still not widespread in France is invited into the debates: the reuse of treated wastewater in indirect drinking water application. This promising practice, already well known in other countries (the United States, China, the Middle East and North Africa), involves reusing water at the outlet of wastewater treatment plants, after treatment, without going through a return in the natural environment, as is usually the case.

The distribution of non-drinking water in urban areas is a fairly old practice that has been experiencing a revival of interest in recent years. Several large cities have developed a dual network of reused water distribution for irrigation and other urban uses. The main categories of urban water reuse are:

– irrigation of green spaces, which includes watering parks, sports fields, greenbelts and residential areas;

– other urban uses such as street cleaning, cleaning of urban furniture, washing of buses and cars, fire protection, air conditioning and supply of water-flushing systems;

– building recycling refers to administrative centers, offices and shopping centers.

An important issue, whether in new developments or already developed areas, is how to integrate water reuse concepts into the environment and how the reused water ultimately reaches the user. Various options are possible, such as distribution via service stations, which allows the reuse water to be transported by cars or trucks, or a separate piping system, which brings the reuse water directly to the user or place of use.

In Japan, wastewater has long been reused in urban areas for non-drinking uses. For example, in the Shinjuku district of Tokyo, treated wastewater is used in 19 buildings for flushing toilets.

Similar wastewater uses are being explored in several pilot projects in Europe and China, mainly as "gray" water treated locally (i.e. all domestic wastewater except toilet water) for flushing toilets and washing machines. Urban reuse for non-drinking purposes (watering of green spaces, waterfalls, fountains, street cleaning, car washing, water flushing and firefighting) is growing rapidly and is a key element in integrated water management policies in heavily populated urban areas.

Figure 3.6. *Non-drinking application: urban furniture cleaning*

Particular attention is paid to possible biological contamination by aerosols, as well as the compliance with microbiological standards for *Cryptosporidium* (< 2 organisms per 10 L), *Giardia* (< 5 organisms per 10 L) and enteroviruses (< 1 organism per 10 L).

In civil engineering, reused water is a good option for construction activities including foundation compaction, dust control, consolidation of embankments and concrete mix.

Some compounds must be reduced or eliminated depending on the application, maintaining the principle of water quality that ensures public safety, as well as the safety of workers and staff.

Fire protection is also one of the alternatives for reused water. Firefighting will require different water qualities depending on the case. Where water can be in direct contact with the staff, the highest quality is required.

Water can be used to supply fire hydrants, sprinkler systems in commercial or industrial facilities or buildings, hotels or motels.

3.1.5.1. *Wooded areas*

In Japan, reused water is pumped to woodlands and ready to provide fire-fighting water. It can also be poured into streams, improving the ecological system in wooded areas and helping to create green belts.

3.1.6. *Drinking water*

Drinking water can be reused directly or indirectly. Countries that have resorted to this practice are facing high pressure on water resources. Social acceptability is the first obstacle to overcome. This practice is actually not well known and may raise questions about health risks. If these questions are legitimate, treatments can reach the different quality classes defined by the legislative framework according to the intended use and to ensure the safety of the practice.

The deployment of the reuse of treated wastewater must therefore be accompanied by a phase of information of the general public. A successful example is Singapore, which uses a large proportion of its treated wastewater for drinking water in a fully accepted way, thanks to the implementation of early awareness-raising activities.

Direct reuse of drinking water is a “pipe-to-pipe” connection between the wastewater treatment plant and the drinking water distribution system. At the outlet of the plant, the water has undergone advanced treatment to meet the strictest standards of drinking water quality.

Indirect reuse of drinking water means that the water at the outlet of the treatment plant after having undergone a high-performance treatment is fed into a reservoir (lake, body of water, dam) or reinjected into groundwater. It is then considered, after pumping, to be available for the production of drinking water and to be part of a water supply program.

The choice of direct or indirect drinking reuse is based on technical factors, such as quality criteria, costs, distribution requirements, treatment techniques and reliability. Other parameters can be considered such as public acceptance.

Indeed, the most serious problem of direct drinking reuse is the acceptance by the public to consume “used” water. Although technological advances make it possible to produce high-quality reused water from urban wastewater, many people fear the direct health effects of ingesting microorganisms (bacteria, viruses, protozoa and helminths) and also organic compounds, such as pesticides and endocrine disruptors. Many scientific studies have shown that there were no relevant objections to the reuse of treated wastewater as drinking water after appropriate treatment.

For other populations, the constraints are religious in nature, with the negative perception of wastewater as unhealthy.

Indirect production of drinking water from wastewater is widespread in the United States, Great Britain and South Africa. The examples of direct production of drinking water of the Windhoek (Namibia) or Krangi (Singapore) plants have never reported any negative influence of this type of water on human health (Davies and Hirji 2003).

Figure 3.7. *Wastewater discharge in river*

Reuse of wastewater is one of many measures that are increasingly being implemented to reduce water deficit or to use drinking water resources for priority uses. There are, today, sufficient examples from around the world of all applications of reused water to demonstrate that this is indeed a viable option that should be considered as an economic and social solution, which meets the many needs of society and people.

3.2. References

Salinity

Belkhodja, M. and Bidai, Y. (2004). Rèponse de la germination des graines d'Atriplex halimus L. sous stress salin. *Sécheresse*, 15(4), 331–335.

Bennaceur, M., Rahmoun, C., Sdiri, H., Medahi, M., Selmi, M. (2001). Effet du stress salin sur la germination, la croissance et la production de grains de blé. *Sécheresse*, 12(3), 167–174.

Cheikh, H., Hamed, R., Abddellaoui, K., Kadri, M., Ben Naceur, S., BelHadj, M. (2008). Évaluation de la tolérance au stress salin de quelque accession d'orge (Hordeum vulgare L.). Cultivées en Tunisie. *Sciences & technologie*, 28, 30–37.

Chen, H. and Jiang, J.G. (2010). Osmotic adjustment and plant adaptation to environmental changes related to drought and salinity. *Environ. Rev.*, 18, 309–319.

El Midaoui, M., Benbella, M., Aït Houssa, A., Ibriz, M., Talouizte, A. (2007). Contribution à l'étude de quelques mécanismes d'adaptation à la salinité chez le tournesol cultivé (*Helianthus annuus* L.). *Hommes, terre et eaux*, 136, 52–63.

Gajardo, H.A., Gomez-Espinoza, O., Ferreira, P.B., Carrer, H., Bravo, L.A. (2023). The potential of CRISPR/Cas technology to enhance crop performance on adverse soil conditions. *Plants*, 12(9), 1892–1915.

Levigneron, A., Lopez, F., Vansuyt, G., Berthomieu, P., Poureroy, P., Casse-Delbart, F. (1995). Les plantes face au stress salin. *Cahier agricultures*, 4, 263–273.

Munns, R. (2002). Comparative physiology of salt and water stress. *Plant, Cell and Environment*, 25, 239–250.

Boron

Cox, F.R. and Reid, P.H. (1964). Calcium-boron nutrition as related to concealed damage in peanuts. *Agron. J.*, 56, 173–176.

Dawson, J.E. and Gustafson, A.F. (1945). A study of techniques for predicting potassium and boron requirements for alfalfa, II. Influence of borax on deficiency symptoms and boron content of the plant and soil. *Soil Sci Soc Amer Proc.*, 10, 147–149.

Eaton, F.M. (1935). Boron in soils and irrigation waters and its effect on peanuts. *USDA Tech Bull.*, 448, 1–132.

Eaton, F.M. (1944). Deficiency, toxicity and accumulation of boron in plants. *Journal Agr. Res.*, 69, 237–279.

Gestring, W.D. and Soltanpour, P.N. (1987). Comparison of soil tests for assessing boron toxicity to alfalfa. *Soil Sci Soc Am J.*, 51, 1214–1219.

Guertal, E.A., Abaye, A.O., Lippert, B.M., Miner, G.S., Gascho, G.J. (1998). Boron uptake and concentration in cotton and soybeans as affected by boron source. *Commun Soil Sci Plant Anal.*, 29, 3007–3014.

Gupta, U.C. (1993). Deficiency, sufficiency, and toxicity levels of boron in crops. In *Boron and Its Role in Crop Production*, Gupta, U.C. (ed.). CRC Press, Boca Raton.

Hanson, E.J., Chaplin, M.H., Breen, P.J. (1985). Movement of foliar applied boron out of leaves and accumulation in flower buds and flower parts of "Italian" prune. *Hort. Sci.*, 20, 747–748.

Kelley, W.P. and Brown, S.M. (1928). Boron in the soils and its relation to citrus and walnut culture. *Hilgardia*, 3(16), 445–458.

Purvis, E.R. and Hanna, W.J. (1940). Vegetable crops affected by boron deficiency in Eastern Virginia. *Virginia Truck Exp. Stat. Bull.*, 105, 69–82.

Stinson, C.H. (1953). Relation of water soluble boron in Illinois soils to boron content of alfalfa. *Soil Sci.*, 75, 31–36.

Warington, K., (1923). The effect of boric acid and borax on the broad bean and certain other plants. *Ann. Bot.*, 37, 629–672.

Wilson, C.M., Louvorn, R.L., Woodhouse, Jr. W.W. (1951). Movement and accumulation of water soluble boron within the soil profile. *Agron J.*, 43, 363–367.

Various applications

Adegoke, A.A., Amoah, I.D., Stenström, T.A., Verbyla, M.E., Mihelcic, J.R. (2018). Epidemiological evidence and health risks associated with agricultural reuse of partially treated and untreated wastewater. A review. *Frontiers in Public Health*, 6, 337–342.

Adewumia, J.R., Ilemobadea, A.A., Vanzyl, J.E. (2010). Treated wastewater reuse in South Africa: Overview, potential, and challenges. *Resources, Conservation and Recycling*, 55, 221–231.

Agoro, M.A., Adeniji, A.O., Adefisoye, M.A., Okoh, O.O. (2020). Heavy metals in wastewater and sewage sludge from selected municipal treatment plants in Eastern Cape Province, South Africa. *Water*, 12, 2746–2758.

Akponikpe, P., Wima, K., Yakouba, H., Mermoud, A. (2011). Reuse of domestic wastewater treated in macrophyte ponds to irrigate tomato in semi-arid West-Africa: Benefits and risks. *Agricultural Water Management*, 98, 834–840.

Alghobar, M.A., Ramachandra, L., Suresha, S. (2014). Effect of sewage water irrigation on soil properties and evaluation of the accumulation of elements in Grass crop in Mysore city, Karnataka, India. *American Journal of Environmental Protection*, 3, 283–291.

Allen, A.J. and Burt, R.G. (2008). Technical concepts related to conservation of irrigation and rainwater in agricultural systems. *Water Resources Research*, 44, 1–16.

Angelakis, A.N. and Snyder, S. (2015) Wastewater treatment and reuse: Past, present, and future. *Water*, 7, 87–95.

Angelakis, A.N., Marecos do Monte, M.H.F., Bontoux, L., Asano, T. (1999). The status of wastewater reuse practice in the Mediterranean basin: Need for guidelines. *Water Research*, 33, 2201–2217.

ANSES (2012). Réutilisation des eaux usées traitées pour l'irrigation des cultures, l'arrosage des espaces verts par aspersion et le lavage des voiries [Online]. Available at: https://www.anses.fr/en/system/files/EAUX2009sa0329Ra.pdf.

Asano, T., Smith, R.G., Tchobanoglous, G. (1985). *Eaux usées municipales : caractéristiques du traitement et de l'eau récupérée. Irrigation avec des eaux usées municipales récupérées – Un manuel d'orientation.* Lewis Publishers, London.

Ayers, R.S. and Westcot, D.W. (1985). *Water Quality for Agriculture. 29 Rev. 1.* FAO, Rome.

Ayers, R.S. and Westcot, D.W. (1994). *Water Quality for Agriculture. 29 Rev. 1.* FAO, Rome.

Aziz, F. and Farissi, M. (2014). Reuse of treated wastewater in agriculture: Solving water deficit problems in arid areas. *Annals of West University of Timişoara, Series of Biology*, 17, 95–110.

Balkhair, K.S. (2016). Impact of treated wastewater on soil hydraulic properties and vegetable crop under irrigation with treated wastewater, field study and statistical analysis. *Journal of Environmental Biology*, 37(5), 1143–1152.

Blumenthal, U.J., Mara, D.D., Peasey, A., Ruiz-Palacios, G., Stott, R. (2000). Guidelines for the microbiological quality of treated wastewater used in agriculture: Recommendations for revising WHO guidelines. *Bull. World Health Organ.*, 78(9), 1104–1116.

Bolong, N., Ismail, A.F., Salim, M.R., Matsuura, T. (2009). A review of the effects of emerging contaminants in wastewater and options for their removal. *Desalination*, 239, 229–246.

Börjesson, S., Melin, S., Matussek, A., Lindgren, P.E. (2009). A seasonal study of the meca gene and Staphylococcus aureus including methicillin-resistant *S. aureus* in a municipal wastewater treatment plant. *Water Research*, 43, 925–932.

Boutin, C., Héduit, A., Helmer, J.M. (2009). Technologies d'épuration en vue d'une réutilisation des eaux usées traitées (REUT). Report, ONEMA, CEMAGREF.

Bouwer, H. and Rice, R.C. (1984). Rénovation des eaux usées au projet d'infiltration rapide de la 23e Avenue. *J. Wat. Pollution. Contrôle Federation*, 56(1), 76–83.

Brissaud, F. (2008). Criteria for water recycling and reuse in the Mediterranean countries. *Desalination*, 218, 24–33.

Caplin, J.L., Hanlon, G.W., Taylor, H.D. (2008). Presence of vancomycin and ampicillin-resistant Enterococcus faecium of epidemic clonal complex-17 in wastewaters from the south coast of England. *Environmental Microbiology*, 10, 885–892.

Carey, C., Lee, H., Trevors, J. (2004). Biology, persistence and detection of *Cryptosporidium parvum* and *Cryptosporidium homynis* oocyst. *Water Research*, 38, 818–868.

Carr, R. (2005). WHO guidelines for safe wastewater use-more than just numbers. *Irrigation and Drainage*, 54, 103–111.

Carre, E. (2017). Qualité biologique des eaux usées traitées en vue de la réutilisation. PhD Thesis, Université de Lyon, Lyon.

Cauchi, A., Hyvrard, N., Nakache, F., Schwartzbrod, J., Zagury, G., Baron, F., Carre, C., Courtois, B., Denis, F., Dernat, P., Larbaigt, G. (1996). Dossier : la réutilisation des eaux usées après épuration. *Techniques, sciences et méthodes*, 2, 81–118.

Chaoua, S., Boussaa, S., Gharmali, A.E., Boumezzough, A. (2019). Impact of irrigation with wastewater on accumulation of heavy metals in soil and crops in the region of Marrakech in Morocco. *Journal of the Saudi Society of Agricultural Sciences*, 18, 429–436.

Chen, W., Lu, S., Pan, N., Jiao, W. (2013). Impacts of long-term reclaimed water irrigation on soil salinity accumulation in urban green land in Beijing. *Water Resources Research*, 49, 7401–7410.

Contreras, J.D., Meza, R., Siebe, C., Rodríguez-Dozal, S., López Vidal, Y.A., Castillo-Rojas, G. (2017). Health risks from exposure to untreated wastewater used for irrigation in the Mezquital Valley, Mexico: A 25-year update. *Water Research*, 123, 834–850.

Crook, J. and Surampalli, R.Y. (1996). Water reclamation and reuse criteria in the U.S. *Water Science and Technology*, 33(10/11), 451–462.

Crook, J., MacDonald, J.A., Trussell, R.R. (1999). Potable use of reclaimed water. *Journal of the American Water Work Association*, 91(8), 40–49.

Davies, R. and Hirji, R. (2003). *Water Resources and Environment. Technical Note F3. Wastewater Reuse*. World Bank, Washington, D.C.

Degrémont, S.A. (2005). *Mémento technique de l'eau*. Degrémont & Suez, Paris.

Dickin, S.K., Schuster-Wallace, C.J., Qadir, M., Pizzacalla, K. (2016). A review of health risks and pathways for exposure to wastewater use in agriculture. *Environmental Health Perspectives*, 124, 900–909.

Ecosse, D. (2001). Techniques alternatives en vue de subvenir à la pénurie d'eau dans le monde. Mémoire D.E.S.S. "Qualité et Gestion de l'Eau", Amiens.

EPA (Environmental Protection Agency) (2002). National recommended water quality criteria: 2002. Report, Office of Water, Office of Science and Technology.

Faby, J.A. (1997). La réutilisation des eaux épurées pour l'irrigation. *Fonds national pour le développement et les adductions d'eau*, 11.

Faby, J.A. and Brissaud, F. (1997). L'utilisation des eaux usées épurées en irrigation. Report, Office International de l'Eau.

Froese, K.L. and Kindzierski, W.B. (1998). Health effects associated with wastewater treatment, disposal, and reuse. *Water Environment Research*, 70(4), 962–968.

Goldstein, R.E.R. (2013). Antibiotic-resistant bacteria in wastewater and potential human exposure through wastewater reuse. PhD Thesis, University of Maryland, Maryland.

Grigg, N., Rogers, P.D., Edmiston, S. (2013). Dual water systems: Characterization and performance for reclaimed water distribution. *Water Research*, 45, 75–106.

Gupta, P. and Diwan, B. (2017). Bacterial Exopolysaccharide mediated heavy metal removal: A review on biosynthesis, mechanism and remediation strategies. *Biotechnology Reports*, 13, 58–71.

Gupta, S.K., Shin, H., Han, D., Hur, H.G., Unno, T. (2018). Metagenomic analysis reveals the prevalence and persistence of antibiotic- and heavy metal-resistance genes in wastewater treatment. *Plant Journal of Microbiology*, 56, 408–415.

Halliwell, D.J., Barlow, K.M., Nash, D.M. (2001). A review of the effects of wastewater sodium on soil physical properties and their implications for irrigation systems. *Australian Journal of Soil Research*, 39, 1259–1267.

Hollander, R., Bullermann, M., Gross, C., Hartung, H., Konig, K., Lucke, F.K., Nolde, E. (1996). Microbiological public health aspects in the use of rain water as water reservoirs for toilet flushing, garden irrigation and laundry. *Gesundheitswesen*, 58(5), 288–293.

Jaramillo, M.F. and Restrepo, I. (2017). Reuse of wastewater in agriculture: A review of its limits and benefits. *Sustainability*, 9, 1734–1741.

Jarup, L. (2003). Hazards of heavy metal contamination. *British Medical Bulletin*, 68, 167–182.

Jeong, H., Kim, H., Jang, J. (2016). Irrigation water quality standards for indirect wastewater reuse in agriculture: A contribution toward sustainable wastewater reuse in South Korea. *Water*, 8, 169–176.

Jiménez, B. and Asano, T. (2008). Water reclamation and reuse around the world. In *Water Reuse: An International Survey of Current Practice, Issues and Needs*, Jiménez, B. and Asano, T. (eds). IWA Publishing, London.

Jindal, A. and Kamat, S. (2011). Water recycling and reuse for domestic and industrial sectors. *Chemical Engineering World*, 32, 52–62.

Jodar-Abellan, A., Lopez-Ortiz, M.I., Melgarejo-Moreno, J. (2019). Wastewater treatment and reuse in Spain. *Water*, 11, 1551–1561.

Junger, J.F. (2000). The reuse of urban waste water in agriculture: European Research on Treated Wastewater Reuse in Agriculture. Report, Urban Technology Network Project, Berlin.

Kalavrouziotis, I.K., Kokkinos, P., Oron, G., Fatone, F., Bolzonella, D., Vatyliotou, M. (2015). Current status in wastewater treatment, reuse and research in some Mediterranean countries. *Desalination and Water Treatment*, 53, 2015–2030.

Kesari, K.K., Soni, R., Sajid, J., Tripathi, P., Lal, J.A., Jha, N.K., Siddiqui, M.H., Kumar, P., Tripathi, V., Ruokolainen, J. (2021). Wastewater treatment and reuse: A review of its applications and health implications. *Water Air Soil Pollution*, 238(208), 56–69.

König, M., Escher, B.I., Neale, P.A., Krauss, M., Hilscherová, K., Novák, J. (2017). Impact of untreated wastewater on a major European river evaluated with a combination of in vitro bioassays and chemical analysis. *Environmental Pollution*, 220(B), 1220–1230.

Lazarova, V., Gaid, A., Rodriguez-Gonzales, J., Alday Ansola, J. (2003). L'intérêt de la réutilisation des eaux usées : analyses d'exemples mondiaux. *Techniques, sciences et méthodes*, 9, 64–85.

Lee, H. and Tan, T.P. (2016). Singapore's experience with reclaimed water: NEWater. *Int. J. Water Resour. Dev.*, 32, 611–621.

Lunn, M. (2001). The deliberate indirect wastewater reuse scheme at Essex & Suffolk Water. In *Colloque de Noirmoutier*, Noirmoutier.

Lyu, S., Chen, W., Zhang, W., Ventilateur, Y., Jiao, W. (2016). Wastewater reclamation and reuse in China: Opportunities and challenges. *J. Environ. Sci.*, 39, 86–96.

Mañas, P., Castro, E., Heras, J. (2009). Irrigation with treated wastewater: Effects on soil, lettuce (*Lactuca sativa* L.) crop and dynamics of microorganisms. *Journal of Environmental Science and Health*, 44, 1261.

Masséna, P.A. (2001). Valorisation des eaux usées en irrigation localisée. Report, Office International de l'Eau.

Monchalin, G. and Aviron-Violet, J. (2000). Réutilisation des eaux usées après traitement. Report, Syndicat National d'Assainissement Autonome et Semi Collectif, Paris.

Njuguna, S.M., Makokha, V.A., Yan, X., Gituru, R.W., Wang, Q., Wang, J. (2019). Health risk assessment by consumption of vegetables irrigated with reclaimed wastewater: A case study in Thika (Kenya). *Journal of Environmental Management*, 231, 576–581.

Okoh, A.I., Sibanda, T., Gusha, S.S. (2010). Inadequately treated wastewater as a source of human enteric viruses in the environment. *International Journal of Environmental Research and Public Health*, 7, 2620–2637.

Oturan, M.A. and Aaron, J.J. (2014). Advanced oxidation processes in water/wastewater treatment: Principles and applications. A review. *Critical Reviews in Environmental Science and Technology*, 44, 2577–2641.

Pedrero, F., Kalavrouziotis, I., Alarcón, J.J., Koukoulakis, P., Asano, T. (2010). Use of treated municipal wastewater in irrigated agriculture. Review of some practices in Spain and Greece. *Agricultural Water Management*, 97, 1233–1241.

PNUE, Tedeschi, S., Pescod, M.B. (1991). Lignes directrices environnementales pour la réutilisation des eaux usées municipales dans la région méditerranéenne. Report, Plan d'action pour la Méditerranée – Programme d'actions prioritaires, Centre d'activités régionales, Split.

Qadir, M., Wichelns, D., Raschid-Sally, L., McCornick, P.G., Drechsel, P., Bahri, A. (2010). The challenges of wastewater irrigation in developing countries. *Agricultural Water Management*, 97, 561–568.

Qadir, M., Dickin, S.K., Schuster-Wallace, C.J., Pizzacalla, K. (2016). A review of health risks and pathways for exposure to wastewater use in agriculture. *Environmental Health Perspectives*, 124, 900–909.

Radcliffe, J.C. and Page, D. (2020). Water reuse and recycling in Australia – History, current status and future prospects. *Water Cycle*, 1, 19–40.

Richards, L.A. (1954). Diagnostic et amélioration des sols salins et alcalins. *Sciences du sol*, 78, 154–155.

Rose, J.B., Farrah, S.R., Friedman, D., Riley, K., Hamann, C.L., Robbins, M. (1999). Public evaluation of advanced water for potable application. *Water Science and Technology*, 40(4/5), 247–252.

Sacks, M. and Bernstein, N. (2011). Utilization of reclaimed wastewater for irrigation of field-grown melons by surface and subsurface drip irrigation. *Israel Journal of Plant Sciences*, 59, 159–169.

Samie, A., Obi, L., Igumbor, O., Momba, B. (2009). Focus on 14 sewage treatment plants in the Mpumalanga province. *Water Air Soil Pollut*, 28, 232–208.

Scott, C.A., Faruqui, N.I., Raschid-Sally, L. (2004). Wastewater use in irrigated agriculture: Management challenges in developing countries. In *Wastewater Use in Irrigated Agriculture: Confronting the Livelihood and Environmental Realities*, Scott, C.A., Faruqui, N.I., Raschid-Sally, L. (eds). CABI Publishing, Wallingford.

Scott, C.A., Drechsel, P., Raschid-Sally, L., Bahri, A., Mara, D., Redwood, M. (2009). Wastewater irrigation and health: Challenges and outlook for mitigating risks in low-income countries. In *Wastewater Irrigation and Health: Assessing and Mitigating Risk in Low-Income Countries*, Drechsel, P., Scott, C.A., Raschid-Sally, L., Redwood, M., Bahri, A. (eds). Earthscan/IWMI/IDRC, Oxford.

Shakir, R., Davis, S., Norrving, B., Grisold, W., Carroll, W.M., Feigin, V., Hachinski, V. (2016). Revising the ICD: Stroke is a brain disease. *Lancet*, 19, 2475–2476.

Sheikh, B., Cooper, R.C., Israel, K.E. (1999). Hygienic evaluation of reclaimed water used to irrigate food crops – A case study. *Water Science and Technology*, 40(4/5), 261–267.

Shoushtarian, F. and Negahban-Azar, M. (2020). Global regulations and guidelines for agricultural water reuse: A critical review. *Water*, 12, 971–992.

Shuval, H.I., Yekutiel, P., Fattal, B. (1985). Preuve épidémiologique de la transmission des helminthes et du choléra par les légumes irrigués avec des eaux usées. Jérusalem – étude de cas. *Science et technologie de l'eau*, 17(4/5), 433–442.

Shuval, H.I., Adin, A., Fattal, B., Rawitz, E., Yekutiel, P. (1986). Irrigation par les eaux usées dans les pays en développement : effets sur la santé et solutions techniques. Technical Document No. 51, World Bank, Washington, D.C.

Shuval, H., Lampert, Y., Fattal, B. (1997). Development of a risk assessment approach for evaluating wastewater reuse standards for agriculture. *Water Science and Technology*, 35(11–12), 15–20.

Singh, A., Sawant, M., Kamble, S.J., Herlekar, M., Starkl, M., Aymerich, E. (2019). Performance evaluation of a decentralized wastewater treatment system in India. *Environmental Science and Pollution*, 26, 21172–21188.

Tacconelli, E., Carrara, E., Savoldi, A., Harbarth, S., Mendelson, M., Monnet, D.L., (2018). Discovery, research, and development of new antibiotics: The WHO priority list of antibiotic-resistant bacteria and tuberculosis. *The Lancet Infectious Diseases*, 18, 318–327.

Tanaka, H., Asano, T., Schroeder, E.D., Tchobanoglous, G. (1998). Estimating the safety of wastewater reclamation and reuse using enteric virus monitoring data. *Water Environment Research*, 70(1), 39–51.

Tortajada, C. and van Rensburg, P. (2020). Drink more recycled wastewater. *Nature*, 577(7788), 26–28.

Tortajada, C., Kastner, M.J., Buurman, J., Biswas, A.K. (2017). California drought: Adaptive responses and building resilience. *Environ. Sci. Politique*, 78, 97–113.

Tripathi, V. and Tripathi, P. (2017). Antibiotic resistance genes: An emerging environmental pollutant. In *Perspectives in Environmental Toxicology*, Kesari, K. (ed.). Springer, Cham, 183–201.

US EPA (2012). Guidelines for Water Reuse 600/R-12/618. Report, Environmental Protection Agency.

Wagner, M. and Bauer, S. (2022). Opportunities and challenges of wastewater reuse – Planning aspects and realized examples. *Water*, 14(10), 1619–1630.

WHO (1989). Health guidelines for the use of wastewater in agriculture and aquaculture. Technical Report No. 74, WHO, Geneva.

WHO (2006). *Guidelines for the Safe Use of Wastewater, Excreta and Greywater: Wastewater Use in Agriculture*. WHO, Geneva.

Winpenny, J., Heinz, I., Koo-Oshima, S., Salgot, M., Collado, J., Hernandex, F. (2010). The wealth of waste: The economics of wastewater use in agriculture. Food and Agriculture Organization of the United Nations.

Xu, Y.B., Hou, M., Li, Y.F., Huang, L., Ruan, J.J., Zheng, L. (2017). Distribution of tetracycline resistance genes and AmpC β-lactamase genes in representative non-urban sewage plants and correlations with treatment processes and heavy metals. *Chemosphere*, 170, 274–281.

Zaman, S.B., Hussain, M.A., Nye, R., Mehta, V., Mamun, K.T., Hossain, N. (2017). A review on antibiotic resistance: Alarm bells are ringing. *Cureus*, 9, 1403–1415.

Zhang, Y. and Shen, Y. (2017). Wastewater irrigation: Past, present, and future. Wastewater treatment: Aims and challenges. *Water*, 6(3), 1234–1245.

4

Regulations

While beneficial in many ways, treated wastewater reuse raises concerns as to potential risks to user health and the environment. In order to reduce significantly the risks, many countries have set clear guidelines to regulate this practice.

4.1. Regulations, state of the art and future challenges of municipal wastewater treatment and water reuse

These guidelines state the criteria and standards to be met for one or more of the reuse steps, which are actually different checkpoints to lower the risk level.

Criteria are established for:

– the effluent quality (monitoring frequency) or minimum treatment to be given to the effluent;

– the design and operation of processing facilities (reserves, failure detection systems, and emergency facilities);

– water distribution and use (type of irrigation, signage for facilities and use, irrigation distances from dwellings and drinking water sources, accessibility to site).

Therefore, the quality aspect is the first required parameter. To enable the use of treated wastewater, special quality requirements are requested so that the use is safe and harmless for humans. In addition, the supply of reused wastewater according to the "adapted to the purpose" principle must take into account different quality requirements so that it can be directly applied for subsequent use.

All of these health risk reduction methods can be used alone or combined. They include chemical and microbiological parameters. As the main health risks associated with the reuse of treated domestic wastewater are microbiological, the quality criteria proposed in this chapter relate to pathogenic microorganisms (viruses, bacteria, protozoa and parasites).

Several countries and international organizations have chosen to frame this practice by regulations or guidelines, including on the quality of the effluent used, the level of treatment to be applied and the measures to reduce exposure to be applied.

For agricultural irrigation, there are several types of regulations: in France (Ministère de la Transition écologique 2022), in Europe (Directives européennes 2020), and WHO recommendations (WHO 2006a).

The main risks associated with the reuse of treated wastewater for irrigation can be classified into three categories:

– health risks;

– agronomic and environmental risks;

– operational risks of degradation of the recycled water quality in distribution systems and of irrigation equipment clogging.

Most of the standards that have emerged in recent years require at least biological treatment of wastewater for reuse in irrigation. In some cases, for example in Mexico, priority has been given to preserving the fertilizing value of wastewater through advanced primary treatment (e.g. coagulation, flocculation and lamellar settling), followed by filtration and/or disinfection for the priority elimination of microbiological pollution, while preserving carbon and nutrients for irrigated crops.

All international regulations set or recommend a level of quality that wastewater treatment must ensure according to the intended use and degree of risk. The degree of risk depends on several factors, including local endemic sanitary conditions, cultivation methods, climatic conditions, risk of direct contact and economic feasibility.

The many regulations that exist in the world vary from country to country. Although these guidelines are constantly evolving, there is now a strong emphasis on separating the reuse of treated wastewater into drinking and non-drinking applications. Reuse for drinking water application generally refers to the increase in the supply of drinking water, whereas reuse for non-drinking application refers to all other uses.

Since not all wastewater treatment processes can completely eliminate wastewater contaminants, reused treated wastewater must meet standards set by the government or relevant institutions working in this field.

The focus is therefore on quality requirements and treatment processes for non-drinking water applications. Only certain countries have standards for reuse to produce drinking water.

This section will review the standards and guidelines of a large number of countries that have experience in this area, as well as the recommendations of the World Health Organization (WHO) and the European Community (EC) Directives.

Quality requirements are a prerequisite for the safe use of reused treated wastewater. Therefore, specific and advanced wastewater treatment processes are essential before reusing the treated wastewater. It should be borne in mind that standards or regulations may vary from country to country. Some countries have standards that specify only one water quality for treated wastewater reuse, while other countries have very detailed standards for each application.

4.2. Regulations for treated wastewater reuse in France

French regulations are strict and binding. In France, it is accepted that the reuse of treated wastewater is a solution for the future in a context of scarcity of water resources. In 2020, the French government set a national target of tripling, by 2025, the volumes of non-conventional water to be reused (treated wastewater, gray water, rainwater, etc.). Therefore, opportunities for the reuse of treated wastewater are encouraged. The *modus vivendi* is: it is essential to reduce water requirements, to open up the possibility of using resources hitherto under-exploited, to reserve water for priority objectives.

The French text (Decree of 10 March 2022) (Ministère de la Transition écologique 2022) responds to the tensions on water resources by reusing treated wastewater. This is only part of the program of the French government that will involve a large number of actions.

The scope of reuse is narrowed with procedural burdens that could nevertheless discourage project leaders. The text provides for a highly regulated reuse to exclude any health risk.

New uses of treated wastewater, previously banned, are now authorized. Only uses that do not lead to the use of treated wastewater within the following places may be authorized:

1) residential use;

2) social, medical and health care establishments and accommodation for the elderly;

3) medical or dental practices, laboratories for biomedical analysis and blood transfusion establishments;

4) pre-schools and primary schools;

5) other institutions open to the public during opening hours.

The following uses are not authorized:

1) food, including drinks, food preparation, cooking and preservation and washing dishes;

2) hygiene of the body and laundry;

3) recreational use, including the use of water for swimming pools and hot tubs, misting, play-water features, decorative fountains accessible to the public.

The Decree of August 29, 2023 makes it possible to consider new urban uses: road cleaning, network flushing (cleaning of pipes with high-pressure waterjet) and fire fighting.

The expansion of uses facilitates the emergence of new multi-purpose projects for the reuse of treated wastewater, making it more economically viable. They are allowed for a limited period of time, in order to advance, through enhanced monitoring, knowledge and data on these new uses. The objective remains in the long term to define national requirements for these uses as is already the case for agricultural irrigation and watering of green spaces, thus facilitating these practices and increasing the volumes of treated wastewater used. The government also committed to experimenting with the use of treated wastewater by food companies.

There are many barriers, particularly related to health and the environment, which explain a certain reluctance. Compared to other countries, some uses of treated wastewater are prohibited: for establishments with a sensitive public (health establishments, pre-schools, schools, etc.), food use, uses related to personal hygiene, recreational uses (swimming pools, fountains, etc.). In addition, the use of treated wastewater is not allowed for health reasons in residential premises, medical and social institutions, early childhood care institutions and other establishments open to the public during opening hours.

Treated wastewater also contains micropollutants. If studies on agricultural irrigation prove reassuring and establish a minimal health risk, these molecules will have to be taken into account for more demanding uses, such as the production of drinking water from treated wastewater, a use still not permitted at present in France.

In addition, a new decree authorizing the water reuse within the agri-food industries is expected and should also help to save significant water amounts.

Figure 4.1. *Drip irrigation. For a color version of this figure, see www.iste.co.uk/gaid/reuse1.zip*

As to the conditions of rainwater use, they may be used for non-domestic uses or in non-food businesses unless this use requires the use of water intended for human consumption. The local and regional authorities have already taken a stand on projects for the reuse of treated wastewater to replace drinking water use, or withdrawals from water courses or groundwater.

A decree should be issued to authorize the use of non-drinking water, such as rainwater, for certain domestic uses (fill the toilet flush, wash floors, laundry, etc.).

The use of treated wastewater will be subject to a prefectural authorization with a limited duration (5 years maximum), for use in the department where it is produced. To this end, the petitioner must also demonstrate to the prefect that their project is compatible with the protection of human health and the environment.

4.2.1. *Application to irrigation*

The following parameters should be taken into account when implementing irrigation. For example, near a sensitive area, special provisions are made.

"Sensitive areas" are those areas that border the plot, outside the irrigated area and within which populations may be exposed to treated wastewater. These include dwellings, courtyards and gardens attached to houses, roads, recreational public spaces and passageways, public buildings and corporate buildings.

Two main categories of information and provisions are distinguished in this document: the steps for processing an application for the approval of a direct reuse project and regulatory values on:

– the health quality level of treated wastewater;

– constraints on the type of area to be irrigated;

– distances to be observed between the recovery area and the areas to be protected. All these provisions confirm the opinion given by AFSSA (French Food Safety Agency) following the proposals of the High Council for Hygiene and Public Health.

Thus, the distinction between four classes of crops is confirmed with the peculiarity, different from the WHO recommendations, of associating type of crop, irrigation methods and number of visitors.

Four levels of sanitary quality of treated wastewater (A, B, C and D) are defined for irrigation:

– A: the authorized crops are vegetable crops for raw consumption, fruit trees and irrigated pastures, spaces open to the public (category A is equivalent to the WHO non-restrictive concept). This level is defined by the bacteriological criterion, that is an *E. coli* content lower than 250 $u \cdot 100\ mL^{-1}$, a reduction higher or equal to 4 log for fecal enterococci, F-specific RNA phages and for spores of sulfite-reducing anaerobes. In addition, the concentration of suspended solids shall not exceed 15 $mg \cdot L^{-1}$ and the COD must be lower than 60 $mg \cdot L^{-1}$. This quality is intended to ensure the protection of consumers of products that can be consumed raw. This quality will also be required to tolerate the watering of sports grounds (golf courses) and green spaces open to the public, subject to compliance with some additional constraints (irrigation outside public opening hours, etc.). In addition to the constraints regarding the microbiological aspect of treated wastewater likely to be reused for irrigation, reused water in this category must also meet chemical quality criteria (presence of chemical micro-pollutants).

– B: the authorized crops are vegetable crops consumed after cooking, cereals and fodder, flower crops, shrubs and nurseries with sprinkling. This level is defined by the bacteriological criterion, that is an *E. coli* content lower than 10^4 $u \cdot 100\ mL^{-1}$, a reduction higher or equal to 3 log for fecal enterococci, F-specific RNA phages and for spores of sulfite-reducing anaerobes. Sprinkler irrigation of these crops is permitted at this quality level provided that the protection of operating staff against the risks of aerosol inhalation is assured, that the sprinkling is carried out at a sufficient distance from dwellings, sports and leisure areas, that obstacles or screens (trees) limiting the spread of aerosols are installed and that direct watering of public roads are avoided.

– C: the authorized crops are grain crops, flower crops, nurseries and fruit trees without sprinkling. This level is defined by the bacteriological criterion, that is an *E. coli* content lower than 10^5 u·100 mL^{-1}, a reduction higher or equal to 2 log for fecal enterococci, F-specific RNA phages and for spores of sulfite-reducing anaerobes.

– D: authorized irrigation: industrial forests with controlled public access. This level is defined by the bacteriological criterion, corresponding to a reduction higher or equal to 2 log for fecal enterococci, F-specific RNA phages and for spores of sulfite-reducing anaerobes.

Treated wastewater is classified in the quality level corresponding to the worst-case parameter.

Reductions are measured between the raw water at the inlet of the wastewater treatment plant and the treated wastewater at the outlet of the wastewater treatment plant or of the additional treatment process, if applicable.

<table>
<tr><th rowspan="2">Parameters</th><th colspan="4">Sanitary quality level of treated wastewater</th></tr>
<tr><th>A</th><th>B</th><th>C</th><th>D</th></tr>
<tr><td>Suspended solids (mg·L⁻¹)</td><td><15</td><td colspan="3" rowspan="2">In compliance with regulations on treated wastewater discharges for the plant outlet outside irrigation period</td></tr>
<tr><td>Chemical oxygen demand (mg·L⁻¹)</td><td><60</td></tr>
<tr><td>Escherichia coli (UFC·100 mL⁻¹)</td><td>≤250</td><td>≤10,000</td><td>≤100,000</td><td>–</td></tr>
<tr><td>Fecal enterococci (reduction in log)</td><td>≥4</td><td>≥3</td><td>≥2</td><td>≥2</td></tr>
<tr><td>F-specific RNA (reduction in log)</td><td>≥4</td><td>≥3</td><td>≥2</td><td>≥2</td></tr>
<tr><td>Spores of sulfite-reducing anaerobes (reduction in log)</td><td>≥4</td><td>≥3</td><td>≥2</td><td>≥2</td></tr>
</table>

Table 4.1. *Sanitary quality levels of treated wastewater (Article 15 of the order of August 2, 2010, France) (Ministère de l'Écologie, de l'Énergie, du Développement durable et de la Mer 2010)*

4.2.1.1. *Use constraints*

Type of use	Sanitary quality level of treated wastewater			
	A	B	C	D
Market gardening, fruit and vegetable crops not processed by a suitable industrial heat treatment, except for watercress crops *(1)*	+	–	–	–
Market gardening, fruit and vegetable crops processed by a suitable industrial heat treatment	+	+	–	–
Pasture crops *(2)*	+	+ *(3)*	–	–
Green spaces open to the public *(4)*	+ *(5)*	–	–	–
Cut flowers	+	+ *(6)*	–	–
Nurseries and shrubs and other flower cultures	+	+	+ *(6)*	–
Fresh fodder	+	+ *(3)*	–	–
Other cereal and fodder crops	+	+	+ *(6)*	–
Fruit farming	+	+ *(7)*	+ *(8)*	–
Short or very short rotation coppice, with controlled public access	+	+	+ *(6)*	+ *(6)*
Forest, excluding short rotation coppice with controlled public access	–	–	–	–

Table 4.2. *Water quality level for each agricultural use (order of August 2, 2010, France) (Ministère de l'Écologie, de l'Énergie, du Développement durable et de la Mer 2010)*

COMMENTS ON TABLE 4.2.–

+: authorized.

–: prohibited.

(1) Reuse of treated wastewater for watercress culture is prohibited.

(2) In the case of sprinkling, animals must not be in the field at the time of the operation and the drinking troughs, if sprayed, must be rinsed before use.

(3) Subject to a time limit after irrigation of 10 days in the absence of a slaughterhouse connected to the wastewater treatment plant, and 21 days otherwise.

(4) Green spaces include motorway service areas, cemeteries, golf courses, racetracks, parks, public gardens, common areas of housing complex, roundabouts and other medians, playgrounds, stadiums, etc. Full, square, stand-up, etc.

(5) Irrigation outside public opening hours or closed to users during irrigation and two hours after irrigation in the case of enclosed green spaces; irrigation during the hours of low attendance and passers-by forbidden access during irrigation and 2 h after irrigation in the case of permanently open green spaces.

(6) Only by localized irrigation.

(7) Prohibited during the period from flowering to picking for unprocessed fruits, except in case of drip irrigation.

(8) Drip irrigation only.

Regarding use constraints, Table 4.2 details the irrigation methods for the series of eleven classes of crops listed. By the "nature of the activities to be protected", the legislator has extended the scope of situations: shellfish farming, swimming and livestock watering are included in the list.

In the case of a greenhouse crop, only localized irrigation is allowed. In case of micro-sprinkling, entry into greenhouses is prohibited during watering and 1 h after.

For suspended solid, chemical oxygen demand and *Escherichia coli* (whose concentration in raw wastewater is relatively constant, around 10^7 cfu·100 mL^{-1}) parameters, these are the limit concentrations to be met at the outlet of wastewater treatment plant.

For fecal enterococci, F-specific RNA phages and spores of sulfite-reducing anaerobes parameters, this is the reduction to be achieved between the inlet and outlet of wastewater treatment plant, helping to assess the efficiency of the treatment system. This ensures that the wastewater treatment provided by the treatment plant is effective against pathogens.

The indicator microorganisms to be monitored reflect the diversity of pathogenic microorganism families present in wastewater and have characteristics that also enable the validation of a wide range of treatments.

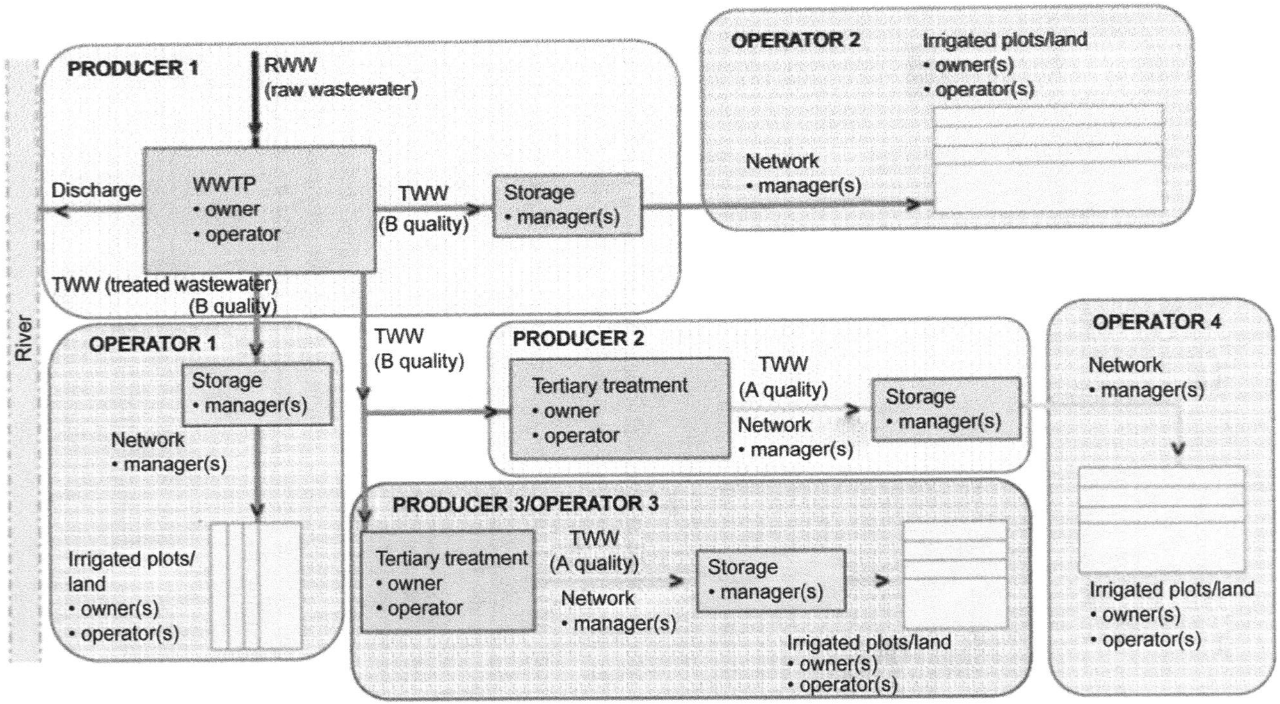

Figure 4.2. *Examples of the division of powers between stakeholders that might be concerned by the implementation of a treated wastewater reuse project (2022 Decree, France) (Ministère de la Transition écologique 2022; WWTP: wastewater treatment plant)*

Green spaces: according to Article 7 of the Decree of December 14, 2023 on conditions for production and use of treated wastewater for watering green spaces, the use of treated wastewater for watering green spaces is prohibited:

1) on water-saturated grounds to prevent runoff of the treated wastewater outside the site;

2) within a close protection perimeter of water collection for human consumption, as defined in Article L1321-2 of the Public Health Code. This prohibition may be waived, after advice from a licensed public health hydrogeologist, in certain areas of the close protection perimeter, in the case of surface water or karst water abstraction;

3) within an area defined by a mayor's or prefect's order, in which the reuse of treated wastewater has a health impact on sensitive water use, such as public abstraction used for human consumption, shellfish farming, fish farming, watercress crop, on-foot fishing, bathing, or nautical activities, and in the absence of a public drinking water network, a well or borehole for domestic water purposes and which has been declared to the mayor of the municipality concerned in accordance with the provisions of Article L2224-9 of the General Code of Local Authorities.

Very precise references (monitoring *E. coli*, sludge, etc.) are issued on the monitoring of wastewater at the outlet of treatment plants and the manager of the plant is clearly assigned to ensure this monitoring and to send information to the Prefect. However, regarding soil monitoring, the responsibility lies with the farmer who is subject to a strict traceability obligation (Figure 4.2), as it is necessary to define the compliance point to specify where the responsibility of the wastewater treatment plant operator stops, the responsibility of the reuse treatment operator if it is different from the previous one and where the responsibility of the next actor in the chain starts. Operators of reuse unit facilities are responsible for the quality of the reused water they deliver to a compliance point. They are responsible for monitoring the quality of the reused water with monitoring programs and frequencies.

4.3. European Directive

The spirit of the European Directive (EC) 2020/41 of May 25, 2020 (European Directives 2020) is to preserve water resources threatened by urban development, human activities and climate change. Its intention is to facilitate the use of treated wastewater reuse whenever it is appropriate and economically efficient as an alternative solution to drinking water for agricultural irrigation (excluding green spaces) by harmonizing common rules applicable in all member states. Quantified targets have been announced: to increase the number of treated wastewater reused from 1.7 billion cubic meters per year to 6.6 billion to achieve a 5% reduction in water stress at EU level by 2025. Although treated wastewater reuse is a promising solution for many member states, at present only a few of them use it and have adopted national legislation or standards in this area.

The EC agrees to introduce minimum levels of harmonization between states, by setting minimum water quality and monitoring requirements. These minimum requirements consist of minimum parameters for reused treated wastewater, based on technical reports from European researchers, and should correspond to the international water reuse standards.

The reuse of treated wastewater for agricultural irrigation can also help promote the circular economy through recovery of nutrients from reclaimed water and its use in crops, using fertigation techniques. Thus, treated wastewater reuse could potentially reduce the need for additional applications of inorganic fertilizers. End users should be informed of the nutrient content of the reclaimed water.

The risk assessment should be based on key elements of risk management and identify any additional water quality requirements necessary to ensure sufficient protection of the environment and human and animal health. To this end, risk management plans for treated wastewater reuse should make sure that the reused treated wastewater is used and managed safely and there are no risks for the environment or human or animal health. Above all, treated wastewater reuse activities must not lead to a deterioration in the quality of water intended for human consumption.

The directive stresses the need to build confidence in treated wastewater reuse by regularly making information available to the public, as well as information and awareness campaigns adapted to the scale of the treated wastewater reuse. Providing clear information on treated wastewater reuse would increase transparency and traceability and ensure public confidence in the use of reused treated wastewater. The provision of information contained in risk management plans and permits will contribute significantly to this. This very good decision increases awareness efforts and promotes the development of the use of treated wastewater.

The treated wastewater reuse uses apply to:

– agricultural irrigation of the following crops:

- food crops consumed raw, that is, crops for human consumption in their raw or unprocessed state,

- processed food crops, that is, crops for human consumption which have been pre-processed (that is, by cooking or industrial processing),

- non-food crops, that is, crops not intended for human consumption (e.g. pasture and fodder, fibers, ornamental crops, seed crops, energy crops, turf crops);

– the reuse of treated wastewater for industrial purposes;

– the use for environmental and community service purposes.

4.3.1. *Minimum requirements applicable to reclaimed water intended for agricultural irrigation*

The directive proposes four classes of quality for reused treated wastewater according to four types of agricultural uses with microbiological thresholds as essential control parameters. The minimum water quality requirements are set out in Table 4.4.

Minimum quality class of reused treated wastewater	Culture category	Irrigation method
A	All food crops consumed raw with the edible part in direct contact with reused water and root crops consumed raw	All irrigation methods
B	Food crops consumed raw with the edible part being cultivated on the surface and not in direct contact with reused water, processed food crops and non-food crops, including crops for feeding milk or meat-producing animals	All irrigation methods
C	Food crops consumed raw with the edible part being cultivated on the surface and not in direct contact with reused water, processed food crops and non-food crops, including crops for feeding milk or meat-producing animals	Drip irrigation or other method of irrigation which makes it possible to avoid direct contact with the edible part of crops
D	Industrial crops, energy crops and seed crops	All irrigation methods

Table 4.3. *Reused treated wastewater quality classes according to uses and agricultural irrigation methods (Directives européennes 2020)*

Crops of a given category (A, B, C or D) shall be irrigated with reused treated wastewater corresponding to the class indicated in Tables 4.3 and 4.4.

The European directive is more stringent than the French requirements for category A. The *E. coli* threshold is 25 times lower ($\leq$10 u·100 mL^{-1}), and BOD5 and SS are required to be ≤10 mg·L^{-1}. Permanent monitoring of turbidity is required to check the treatment effectiveness. For B quality, the threshold is even more strict (factor 100), and there is same requirement for C quality water.

Legionella bacteria and intestinal nematodes are still requested in analytical reviews. On the other hand, health compliance checks will be more frequent, depending on the quality class of the reused treated wastewater, so that problems can be responded to faster.

<table>
<tr><th rowspan="2">Reclaimed water quality class</th><th rowspan="2">Indicative technological objective</th><th colspan="5">Quality requirements</th></tr>
<tr><th>E. coli u·100 mL^{-1}</th><th>BOD5 mg·L^{-1}</th><th>SS mg·L^{-1}</th><th>Turbidity NTU</th><th>Other</th></tr>
<tr><td>A</td><td>Secondary treatment, filtration and disinfection</td><td>≤10</td><td>≤10</td><td>≤10</td><td>≤5</td><td rowspan="4">Legionella spp.: <1,000 UFC·L^{-1} where there is a risk of aerosol formation

Intestinal nematodes (helminth eggs): ≤1 egg/L for pasture and fodder irrigation</td></tr>
<tr><td>B</td><td>Secondary treatment and disinfection</td><td>≤100</td><td rowspan="3">25</td><td rowspan="3">35</td><td>–</td></tr>
<tr><td>C</td><td>Secondary treatment and disinfection</td><td>≤1,000</td><td>–</td></tr>
<tr><td>D</td><td>Secondary treatment and disinfection</td><td>≤10,000</td><td>–</td></tr>
</table>

Table 4.4. *Minimum quality requirements applicable to reused water intended for agricultural irrigation (Directives européennes 2020)*

The operator of the wastewater treatment plant must have a comprehensive risk management plan validated to get an operating permit. It includes a list of the risks and indicates the means implemented to manage them. This enables anticipating possible incidents and to make the use of water safe. Operators of treated wastewater reuse facilities conduct routine monitoring to check that the reused treated wastewater meets minimum water quality requirements. This routine monitoring is part of the treated wastewater reuse system verification procedures. Risk assessment and management is at the heart of the EU directive, which introduces the concept of barrier to compensate for the strictness of quality criteria. In practice, if you add a barrier (different treatment stages, distances related to an activity, period of a few days between irrigation and the return of animals, etc.), quality criteria are less stringent, enabling the use of B-quality water while quality A is required. The directive provides for the possibility of adapting the system to the characteristics of the project, on a case-by-case basis.

Validation monitoring involves checking indicator microorganisms associated with each group of pathogens, namely, bacteria, viruses and protozoa. The selected indicator microorganisms are *E. coli* for pathogenic bacteria, F-specific coliphages, somatic coliphages or coliphages for pathogenic viruses, and *Clostridium perfringens* spores or sulfite-reducing anaerobes and their spores for protozoa. The performance targets (10-log reduction) for the validation monitoring of the selected indicator microorganisms are listed in Table 4.5 and must be reached at the compliance point, taking into account the concentrations of raw wastewater entering the urban wastewater treatment plant. At least 90% of samples taken for validation meet or exceed performance targets.

Reclaimed water quality class	Indicator microorganisms (*4)	Performance targets of the treatment chain (10-log reduction)
A	*E. coli*	≥5.0
	Total coliphages F-specific coliphages Somatic coliphages (*5)	≥6.0
	Clostridium perfringens spores Sulfite-reducing anaerobes and their spores (*6)	≥4.0 (in the case of the *Clostridium perfringens* spores) ≥5.0 (in the case of sulfite-reducing anaerobes and their spores)

Table 4.5. *Validation monitoring of reclaimed water intended for agricultural irrigation (European Directives 2020)*

COMMENTS ON TABLE 4.5.–

*(*4) The reference pathogens Campylobacter, rotavirus and* Cryptosporidium *may also be used for validation monitoring instead of the proposed indicator microorganisms. The following performance targets, expressed as 10-log reduction, shall apply in this case: Campylobacter (≥5.0), rotavirus (≥6.0) and* Cryptosporidium *(≥5.0).*

*(*5) Total coliphages are chosen as the most appropriate viral indicator. However, if the analysis of total coliphages is not possible, at least one of them (F-specific coliphages or somatic coliphages) must be analyzed.*

*(*6)* Clostridium perfringens *spores are chosen as the most appropriate protozoa indicator. However, sulfite-reducing anaerobes and their spores offer an alternative solution if the concentration of* Clostridium perfringens *spores makes it impossible to validate the 10-log reduction required.*

If a biological indicator is not present in sufficient quantity in the raw wastewater to achieve a 10-log reduction, the absence of this biological indicator in reclaimed water means that the validation requirements are met. Compliance with the performance target can be determined through analytical control, adding up the performance assigned to each treatment step based on scientific evidence for well-established conventional methods, such as published data from trial reports or case studies, or based on laboratory tests under controlled conditions for innovative treatments.

The risk management plan becomes the central pivot of each project, also setting the governance of the responsibilities of all stakeholders involved, from the producer of reused water to the end-user.

4.4. World Health Organization

The WHO was the first international organization to issue a directive for treated wastewater reuse in agriculture, followed by the FAO and the World Bank. In 2012, the WHO completely revised its 1989 and 2006 recommendations for agricultural irrigation with treated wastewater. Guide values have been maintained at a level which is thus able to limit health risks to a level acceptable for the protection of public health, but with new concepts of additional barriers, which ensures the economic viability of reuse projects.

Indeed, the WHO has identified four measures to reduce the risk of treated wastewater reuse on public health, including:

– wastewater treatment;

– crop limitation;

– the control of treated wastewater use;

– exposure control.

Therefore, the WHO has developed guidelines that take into account these four measures for the proper reuse of treated wastewater in agriculture. They are called as guidelines for the microbiological quality of treated wastewater used in agriculture.

These guidelines set out the number of coliform bacteria considered as indicators of pathogens and the number of nematode eggs acceptable in a final effluent that may be intended for irrigation. These numbers are considered as standards. Their main objective is to eliminate health risks so that irrigation can be carried out without restriction. These standards define the quality of treated wastewater for irrigation. They were established to protect users and farm workers, protect consumers of agricultural products, protect surface and ground water resources and soils, protect irrigation equipment and maintain acceptable yields.

The WHO texts consider several types of irrigation:

– *Submersion and infiltration irrigation*: the use of infiltration and submersion irrigation techniques poses the greatest risks to farm workers and their families. The risk is particularly high when these people move land by hand (Blumenthal et al. 2000; Blumenthal 2001). Nevertheless, wastewater treatment with a reduction of pathogens of 2–3 logarithmic units protects agricultural workers.

– *Spray and sprinkler irrigation*: spray and sprinkler irrigation have the greatest potential for spreading contamination to surface crops and for harming communities living nearby. Bacteria and viruses (but usually not helminth eggs or protozoan cysts and oocysts) can be transmitted through aerosols to neighboring communities.

– *Local irrigation*: local irrigation techniques (such as nozzle or drip irrigation) offer the best health protection for agricultural workers, as wastewater is applied directly to plants.

The 2012 edition presents an updated text including all scientific knowledge on chemical and microbiological parameters related to health risks. For the microbiological part, the WHO used as dose-response model a β-Fish model for rotavirus and *Campylobacter* infections and an exponential model for *Cryptosporidium* infections (Fattal et al. 2004; Gauvin 2006). Models are governed by the following equations:

a) dose-response β-Fish model (for *Campylobacter* and rotaviruses):

$$PI(d) = 1 - [1 + (d/DI50)\,(2^{1/\alpha} - 1)]^{-\alpha}$$

b) exponential dose-response model (for *Cryptosporidium*):

$$PI(d) = 1 - Exp\,(-\,rd)$$

c) infection annual risk:

$$PI(A)(d) = 1 - [1 - PI(d)]^{n}$$

where:

– PI(d) is the risk of infection for an individual exposed (by ingestion in this case) to a single pathogenic dose d;

– PI(A)(d) is the annual risk of infection for an individual resulting from n exposures per year, at single pathogenic dose d;

– DI50 is the median infectious dose;

– α and r are the infectivity constants of the pathogen;

– for rotaviruses, DI50 = 6.17 and $\alpha = 0.253$;

– for *Campylobacter*, DI50 = 896 and $\alpha = 0.145$;

– for *Cryptosporidium*, r = 0.0042 (Fattal et al. 2004; Gauvin 2006);

– PI(A)(d) can also be interpreted as the risk over a shorter (or longer) period – for example, monthly risk, with n designating in this case the number of exposures during m months. The value of PI(A)(d) is between 0 and 1. If PI(A)(d) = 1, infection is certain.

It appears that the health risks associated with chemicals in wastewater require a great deal of attention, although there is very little evidence of direct health impacts from exposure to chemicals resulting from the use of wastewater in agriculture.

The implementation of inappropriate practices for irrigation with untreated or partially treated water also has a negative impact on the quality and safety of groundwater from shallow aquifers and surface water that can be used as a drinking water supply. Nitrate contamination of aquifers is an example of wastewater poorly treated and reused without control.

The use of treated wastewater can lead to the penetration of potentially toxic pollutants into soils. These toxic pollutants can be harmful to health and well-being by transferring through the food chain, due to the absorption of chemicals from soils by plants. Pollutants accumulated in soils as a result of irrigation with wastewater can also subsequently contaminate surface water and groundwater. Organic chemicals are eliminated or decomposed during the treatment of wastewater. The results indicate that the frequency of detection of most of these organic chemicals is lower than 10%. This is why the WHO considers that there is no real need to take them into account for the use of wastewater in agriculture. However, if inadequately treated water is frequently used to irrigate crops, these chemicals should be included in the assessment.

Emerging substances (hormones, pesticides, endocrine disruptors, persistent organic pollutants, non-ionic detergents and human or veterinary pharmaceutical residues) are present in treated wastewater at various concentrations (see Chapter 2). There is a lack of direct evidence of adverse human health effects from reused treated wastewater, and the WHO stresses that studies on these substances are necessary to assess their potential health and environmental risks in the context of the use of treated wastewater to irrigate crops. Based on the limited data provided by publications, they appear to be highly adsorbed by soil organic matter and therefore have little likelihood of accumulating in harvested plants at concentrations that present a health risk in case of consumption. No toxic effects on humans were observed from exposure to these chemicals in the context of treated wastewater use in agriculture.

4.4.1. *Objectives of pathogen reduction for the different irrigation types*

4.4.1.1. *Unrestricted irrigation*

The results of the model application for unrestricted irrigation, combined with relevant epidemiological evidence, show that to reach $\leq 10^{-6}$ DALY per person per year for rotaviruses, a total pathogen reduction of 6 logarithmic units must be achieved before consuming leaf crops (lettuces) and seven logarithmic units to consume root crops (onions) (WHO 2012).

In these guidelines, the performance target used to achieve a tolerable additional disease burden $\leq 10^{-6}$ DALY per person per year in the case of irrigation without restriction is a reduction of pathogens by six to seven logarithmic units. As the risks associated with exposure to rotavirus are considered to be the most serious, this level of pathogen reduction provides sufficient protection against infection by bacteria or protozoa. It is possible to achieve a reduction of pathogens by six to seven logarithmic units by applying appropriate health protection measures, each of which is associated with a reduction or range of reductions in logarithmic units (WHO 2012).

In the case of irrigation, the required quality of effluents will depend on the crop(s) to be irrigated, soil conditions and the adopted effluent distribution system. By restricting crops and selecting irrigation systems that minimize health risks, the degree of wastewater treatment before application can be reduced. The most appropriate wastewater treatment to be applied before using effluents in agriculture is one that will produce an effluent that meets the microbiological and chemical quality recommendations and the minimum operating and maintenance requirements.

Microbial reduction targets to protect consumers from helminth infestation are based on the results of epidemiological and microbiological studies. The value of ≤1 helminth egg per liter is confirmed in the case of unrestricted irrigation.

Unrestricted irrigation is divided into five categories:

– Class A: shows that the required pathogen reduction can be achieved by combining wastewater treatment ensuring a pathogen reduction by four logarithmic units, a pathogen decay between the last irrigation and consumption (reduction by two logarithmic units) and washing with water salads or vegetables before consumption (reduction by one logarithmic unit). This option, which reduces pathogens by seven logarithmic units, is suitable when root crops that can be consumed raw are irrigated with treated wastewater.

– Class B: provides for less intensive wastewater treatment than class A (three logarithmic units instead of four), combined with two health protection post-treatment measures: a reduction by two logarithmic units per decay and a reduction of one logarithmic unit by washing salads or vegetables with water before consumption. This class, which provides a pathogen reduction of six logarithmic units, is suitable for irrigation of crops other than roots (salad type) and vegetables that will be consumed raw. It provides a sufficient level of health protection, as salads are often less contaminated than root crops and therefore their consumption is less risky.

– Class C: provides for an even less intensive treatment (two logarithmic units), but in combination with drip irrigation of tall crops (such as tomatoes), which makes it possible to obtain the remaining required reduction of four logarithmic units.

– Class D: includes drip irrigation of low-lying crops other than roots and tubers (reduction by two logarithmic units) and therefore must include further treatment (four logarithmic units).

– Class E: relies solely on wastewater treatment to achieve the required six to seven logarithmic units reduction.

To sum up:

– Class A: suitable for root vegetables. 7-log pathogen reduction obtained by the treatment combination (4 log) + preharvest interval + clean water washing.

– Class B: suitable for leafy vegetables. 6-log pathogen reduction obtained by treatment combination (3 log) + preharvest interval + clean water washing.

– Class C: suitable for tall crops and aboveground production, without picking products from the ground. 6-log reduction obtained by treatment combination (2 log) + drip irrigation.

– Class D: suitable for low-lying crops and aboveground production, without picking products from the ground. 6-log reduction obtained by the treatment combination (4 log) + drip irrigation.

– Class E: suitable for all low-lying crops. 7-log reduction by treatment.

A typical sequence of wastewater treatment processes to achieve this reduction includes conventional wastewater treatment (primary sedimentation, activated sludge, including secondary sedimentation, for example), followed by chemical coagulation, flocculation, sedimentation and disinfection (chlorination or irradiation by UV). Such a sequence is used, for example, in California to ensure that this state's criteria for water recycling are met in the case of unrestricted irrigation (≤2.2 total coliforms per 100 mL and turbidity ≤2 nephelometric units) (Cooper et al. 2012). This option does not take into account the reduction of pathogens due to natural decay between the last irrigation and consumption and some household food preparation practices, such as washing, disinfection, peeling and/or cooking. The overall health protection obtained is therefore even higher than 10^{-6} DALY per person per year.

Class	Required pathogen reduction (log units)	Monitoring/verification level (*E. coli* for 100 mL)	Applications
A	4	$\leq 10^3$	Root crops
B	3	$\leq 10^4$	Leafy crops
C	2	$\leq 10^5$	Drip irrigation of tall crops
D	4	$\leq 10^3$	Drip irrigation of low-lying crops
E	6–7	1 or ≤10	Verification level depends on the local regulation agency requirements[a]

[a]For example, in the case of secondary treatment, followed by filtration and disinfection: BOD5 < 10 $mg \cdot L^{-1}$; turbidity < 2 NTU; residual chlorine: 1 $mg \cdot L^{-1}$; pH: 6–9; fecal coliforms: undetectable for 100 mL.

Table 4.6. *Wastewater treatment monitoring/verification (number of* E. coli *for 100 mL of treated wastewater) for various degrees of wastewater treatment in the context of unrestricted irrigation*

The guidelines also imply that protozoan cysts should be reduced to the same level as helminth eggs.

4.4.1.2. *Restricted irrigation*

The risk in this case concerns workers in direct contact with wastewater. In the case of restricted crop irrigation, the health target of $\leq 10^{-6}$ DALY per person per year for rotavirus requires a reduction in the number of *E. coli* by four logarithmic units from 10^7–10^8 per 100 mL in raw water to 10^3–10^4 per 100 mL in treated water.

Performance target for restricted irrigation is set at ≤1 helminth egg per liter of treated wastewater.

Class	Required pathogen reduction (log units)	Monitoring/verification level (*E. coli* for 100 mL)	Applications
F	4	$\leq 10^4$	Labor-intensive agriculture
G	3	$\leq 10^5$	Highly mechanized agriculture
H	0,5	$\leq 10^6$	Elimination of pathogens in septic tank

Table 4.7. *Wastewater treatment monitoring/verification (number of* E. coli *for 100 mL of treated wastewater) for various degrees of wastewater treatment in the context of restricted irrigation*

Options F, G and H apply to restricted irrigation:

– Class F: refers to labor-intensive agriculture, and the health target is achieved by a reduction of four logarithmic units in pathogens. Wastewater must be treated to reach an *E. coli* concentration $\leq 10^4$ per 100 mL.

– Class G: refers to highly mechanized agriculture, and the health target is achieved by a reduction of three logarithmic units in pathogens. Wastewater must be treated to reach an *E. coli* concentration $\leq 10^5$ per 100 mL.

– Class H: is typical for an isolated household or establishment: minimum treatment in a septic tank (reduction of pathogens by 0.5 logarithmic units). The first stage of the project is to be carried out in a semi-underground system, followed by underground irrigation through a system absorbing the septic tank effluents from the soil.

There is no contact between crops and pathogens present in the septic tank effluents, so that the remaining reduction of pathogens by 6.5 logarithmic units for root crops can be allocated to the underground irrigation system.

To sum up:

– Class F: it includes intensive manual labor agriculture. 4-log reduction is obtained by treatment, whereas 3-log reduction is ensured by crop restriction.

– Class G: it includes highly mechanized agriculture. 3-log reduction is obtained by treatment, whereas 4-log reduction is ensured by the restriction of crops.

– Class H: autonomous sanitation or equivalent: it includes septic tank (1 log) followed by underground spreading (6 log).

For helminth eggs, Table 4.8 summarizes the threshold values to be obtained for the different uses of classes A to G according to the uses.

	Raw water concentration ($u \cdot L^{-1}$)	**Required reduction (log unit)**	**Threshold value ($u \cdot L^{-1}$)**	**Comments**
Treatment	1,000	3	≤1	Design according to the input concentration and its variation
	100	2	≤1	
	10	1	≤1	
	≤1	0	–	–
Treatment + products washing	1,000	2	≤1	Treatment and disinfection through washing
	100	1	≤1	
	10	0	–	Single washing
	≤1	0	–	–

Table 4.8. *Reduction of helminth eggs according to applications and required threshold values*

4.4.1.3. *Local irrigation*

Where the implemented wastewater application technique is local drip irrigation, reductions in logarithmic units of pathogens as shown in Table 4.6 for classes A to E should be used.

For low-lying crops, the microbial reduction target aiming for ≤1 helminth egg per liter of treated wastewater should be applied. However, when local irrigation is used to irrigate tall crops (i.e. crops whose harvested parts are not in contact with the soil), performance targets for helminth egg concentrations are not applicable.

4.4.2. *Monitoring/verification*

Health targets are met by three types of monitoring:

– Validation proves that the system as a whole meets performance targets.

– Operational monitoring through all analyses and tests is measured quickly to anticipate risk management decisions, including bacterial (*E. coli*) indicators and helminth eggs, as well as indicators such as BOD5 and SS.

Monitoring/verification is performed periodically to demonstrate that the system is functioning properly. It applies to both local and unrestricted and restricted irrigation.

4.5. Regulation on treated wastewater reuse in the United States

The United States Environmental Protection Agency (US EPA) does not require or restrict any type of treated wastewater reuse. The United States is considered the world leader in treated wastewater reuse, with the first regulation adopted in 1918 in California, followed by several revisions, the most recent date being December 2000. At the federal level, the US EPA revised its 2004 recommendations in 2012, mainly maintaining two guide levels for recycled water quality and recommendations for better monitoring of treatment processes (U.S. Environmental Protection Agency and U.S. Agency for International Development 2004; Shoushtarian and Negahban-Azar 2020).

States generally retain primary regulatory authority (i.e. primacy) in the allocation and development of water resources. Some states have put in place programs to deal specifically with reuse, and some have integrated treated wastewater reuse into their existing programs. The EPA, states, tribes and local governments implement programs under the Safe Drinking Water Act (SDWA) and the Clean Water Act (CWA) to protect the quality of drinking water resources, community drinking water and water bodies such as rivers and lakes. Together, the two directives provide a basis from which states can authorize, regulate and supervise water use as they see fit.

Among others, the US EPA published guidelines for treated wastewater reuse in 2012. These are based on the zero fecal indicator approach, that is, they aim for no fecal bacterial indicator such as total, fecal coliforms or *E. coli,* in treated wastewater reused (U.S. Environmental Protection Agency and U.S. Agency for International Development 2004, 2012).

Applications	Treatment	Reused treated wastewater quality	Monitoring/ analyses
Urban uses			
Unrestricted: ***treated wastewater*** is reused for non-drinking uses where public access is not restricted	Secondary Filtration Disinfection	pH 6.0–9.0 Turbidity ≤ 2 NTU BOD5 ≤ 10 mg·L^{-1} Fecal coliforms: undetectable in 100 mL 1 mg·L^{-1} Cl_2 residual	pH: weekly BOD5: weekly Turbidity: continuously Fecal coli: daily Residual Cl_2: continuously
Restricted: ***treated wastewater*** is reused for non-drinking uses where public access is controlled or restricted by fences, signs or temporary access	Secondary Disinfection	pH: 6.0–9.0 BOD5 ≤ 30 mg·L^{-1} (Australian Academy of Technological Sciences and Engineering 2004) SS ≤ 30 mg·L^{-1} Fecal coli. ≤ 200 for 100 mL (Blumenthal et al. 2000; Blumenthal 2001; Armon 2002) 1 mg·L^{-1} Cl_2 residual (minimum) (Ayres 1992)	pH: weekly BOD5: weekly SS: daily Fecal coli: daily Residual Cl_2: continuously
Agriculture			
Food crops Water is reused for surface or sprinkler irrigation for food intended to be consumed raw	Secondary Filtration Disinfection	pH: 6.0–9.0 BOD5 ≤ 10 mg·L^{-1} (Australian Academy of Technological Sciences and Engineering 2004) Turbidity ≤ 2 NTU (Al-Jasser 2011) Fecal coliforms: undetectable for 100 mL (Armon 2002) 1 mg·L^{-1} Cl_2 residual (minimum) (Ayres 1992)	pH: weekly BOD5: weekly Turbidity: continuously Fecal coli: daily Residual Cl_2: continuously

Industrial food crops Water is reused for surface or sprinkler irrigation for food intended to be consumed after industrial processing Non-food crops Water is reused for the irrigation crops not consumed by humans, including fodder, fibers and seed crops, as well as pasture and nursery irrigation	Secondary Disinfection	pH: 6.0–9.0 BDO5 $\leq$ 30 $mg\cdot L^{-1}$ (Australian Academy of Technological Sciences and Engineering 2004) SS $\leq$ 30 $mg\cdot L^{-1}$ Fecal coli. $\leq$ 200 for 100 mL (Blumenthal et al. 2000; Blumenthal 2001; Armon 2002) 1 $mg\cdot L^{-1}$ Cl_2 residual (minimum) (Ayres 1992)	pH: weekly BOD5: weekly SS: daily Fecal coli: daily Residual Cl_2: continuously

Table 4.9. *Recommendations for various reused water uses of the US EPA (U.S. Environmental Protection Agency and U.S. Agency for International Development 2004. 2012)*

Applications	Arizona	Florida	Hawaii	Nevada	Texas
Unrestricted urban	Undetectable (average)	75% of samples undetected	2.2.100 mL^{-1} (average)	2.2.100 mL^{-1} (average)	2.2.100 mL^{-1} (average)
	23.100 mL^{-1} (maximum)	25.100 mL^{-1} (maximum)	23.100 mL^{-1} (max. in 30 d)	23.100 mL^{-1} (maximum)	75.100 mL^{-1} (maximum)
Controlled and restricted urban	200.100 mL^{-1} (average)	75% of samples undetected	23.100 mL^{-1} (average)	23.100 mL^{-1} (average)	200.100 mL^{-1} (average)
	800.100 mL^{-1} (maximum)	25.100 mL^{-1} (maximum)	200.100 mL^{-1} (maximum)	240.100 mL^{-1} (maximum)	800.100 mL^{-1} (maximum)
Unrestricted water treatment	Secondary treatment Filtration Disinfection	Secondary treatment Filtration Reinforced disinfection	Secondary treatment Oxidation Filtration Disinfection	Secondary treatment Disinfection	Unspecified

Table 4.10. *Criteria selected by some states for fecal coliforms (UFC.100 mL^{-1}) for wastewater reuse in urban environment with and without restriction (U.S. Environmental Protection Agency and U.S. Agency for International Development 2004)*

The established criterion is 0 fecal coliform.100 mL^{-1} (average) and should never exceed 14 fecal coliforms.100 mL^{-1}. The EPA guidelines suggest a treatment method, criteria for microbiological, physicochemical and aesthetic quality of water, a monitoring frequency and parameters, as well as a buffer zone, for the various possible uses of wastewater. Regarding the microbiological quality criterion for urban uses such as irrigation of all types of land (parks, golf courses, cemeteries), it is 0 fecal coliform per 100 mL (average) and should never exceed 14 fecal coliforms per 100 mL. Under certain very special conditions (forbidden, restricted or not frequent access, example: silviculture site), an average of 200 fecal coliforms per 100 mL can be accepted without ever exceeding 800 fecal coliforms per 100 mL. The EPA has adopted the zero fecal indicator approach. Regarding viruses, it is mentioned that their presence in wastewater is a concern. However, it is explained that limits are not specified because their identification and quantification are technically difficult and expensive to monitor. In addition, the authors of these recommendations state that there is no consensus on the health effects of small amounts of viruses and that no cases of viral infection associated with treated wastewater reuse were ever reported in the United States. However, a complete microbiological characterization, without further details on the pathogenic microorganisms sought, must be carried out before the start of activities at a point of use (Huot 2008).

Washington, California and Florida have developed comprehensive regulations on landscape irrigation. Details of the regulations of each state can be found in the previous studies (California Code of Regulations 2018; Washington State Department of Ecology 2018; California State Water Resources Control Board 2020; Florida Department of Environmental Protection 2021). Washington standards define that 13 of C-class or higher reclaimed waters can be used in restricted areas and that A-class or higher reclaimed water can be used in open access areas. The classification of the reclaimed water is based on disinfection and treatment requirements, and these requirements can be found in the reference (Washington State Department of Ecology 2018). California has similar regulations for landscape irrigation, which require two different qualities of reclaimed water based on public access to the area of use. However, the reclaimed water used in public access areas must meet stricter requirements, such as turbidity, chlorine contact time after chlorination, the F-specific bacteriophage MS2 in addition to disinfection requirements (California Code of Regulations 2018).

Florida regulates the quality of reclaimed water based on public access to the area of use. The requirement for TSS concentration prior to disinfection is included in regulations along with process and disinfection requirements.

Ornamental reuse of treated wastewater can be aesthetic, such as decorative pools, decorative fountains, ponds, lagoons and landscape retention ponds. In Florida, water reclaimed for these purposes is required to meet the same high-quality standards, regardless of purpose. However, Washington has different requirements for the qualities of water reclaimed for landscape dams and decorative fountains, respectively. Since the reclaimed water in decorative fountains is more likely to be in contact with humans than water in landscape dams, the highest quality of reclaimed water is required for decorative fountains.

Agricultural irrigation using reclaimed water to support agricultural production has been a well-established practice for years in arid and semi-arid regions. A significant part of existing treated wastewater reuse systems provides reclaimed water for agricultural irrigation. In California, about 63% of the total volume of reclaimed water is used for agricultural irrigation. In Florida, about 34% of the total reclaimed water is used for agricultural irrigation.

California regulations distinguish two different types of recycled water: non-drinking and drinking applications. The non-drinking regulations are divided into four different treatment levels, referring to the respective application. The California Code of Regulations (CCR) specifies the following types of uses for recycled water: for irrigation, dams and cooling for other purposes. Applications for the use of recycled drinking water are managed by the California Water Code and are divided into "direct drinking reuse", "indirect drinking reuse for groundwater recharge" and "increase in reservoir water" (California Code of Regulations 2018).

4.6. Regulation on water reuse in China

China has published very detailed standards on treated wastewater reuse. To support the development of treated wastewater reuse, the Chinese government has enacted a series of regulations for a variety of reuse applications for about 20 years. These are the Standard for Environment Reuse (GBT 18921-2002), the Standard for Miscellaneous Urban Reuse (GBT 18920-2002), the Standard for Industrial Reuse (GBT 19923-2005), the Standard for Farmland Irrigation Reuse (GB20922-2007) and the Standard for Green Space Irrigation Reuse (GB/T 25499-2010) (Water Unit Environmental Health Directorate 2011).

Therefore, all wastewater treatment plants producing reused treated wastewater have specific treatment steps to meet the requirements of selected uses.

4.7. Australia

The Environmental Protection and Heritage Council and the Natural Resource Management Ministerial Council have developed new national guidelines, published in October 2005. In 2011, guide values for the treated wastewater reuse for non-drinking purposes were published for the regions of Western Australia (Water Unit Environmental Health Directorate 2011).

They propose a risk management framework based on:

– the analysis of the different risks inherent in the reuse of treated wastewater;

– the control at the recycling stages where risks are highest (hazard analysis and critical control point);

– monitoring of parameters and risk tracking.

Figure 4.3. *Reused water for environmental use (Australia). For a color version of this figure, see www.iste.co.uk/gaid/reuse1.zip*

The different Australian states and territories have published their own recommendations, which differ from each other, leading to heterogeneity in this area and public understanding. As such, entities such as South Australia (SA), Western Australia (WA), New South Wales (NSW), Australian Capital Territory (ACT), Victoria Queensland (VIC QLD) requested a common text for the guidelines. As everywhere, health risks correspond most often to microbiological quality criteria. This risk has been set at 10^{-6} DALY, based on the WHO drinking water quality recommendations. Risk control measures are aimed at reducing the concentration of pathogenic micro-organisms in water by preventing their introduction at the source or by treating water. Reducing people's exposure is recommended by controlling use (type of use, irrigation technique, signage, etc.). Limits on the quantity of pathogenic micro-organisms present in treated waters are set out in these guidelines, for different uses and depending on the other risk reduction measures applied (Table 4.11). In addition to a quality level in relation to microbiological and physicochemical indicators, some standards also recommend treatment lines and additional barriers to reduce health risks.

Use	Treatment	Reused treated water quality
Municipal, ornamental irrigation **Uncontrolled public access**	Secondary + filtration, disinfection	Thermotolerant coliforms <10 UFC.100 mL^{-1} with four values out of five <40 UFC.100 mL^{-1} 1 mg·L^{-1} of residual chlorine after 30 min pH: 6.5–8.0
Municipal, ornamental irrigation **Controlled public access** **Horticulture** **Pasture and fodder**	Secondary + disinfection or lagooning Secondary Secondary	Thermotolerant coliforms <1,000 UFC.100 mL^{-1} with four values out of five <4,000 UFC.100 mL^{-1}
Garden watering **Toilet flushing** **Car washing** **Path/wall cleaning**	Secondary + filtration, disinfection	Thermotolerant coliforms <10 UFC.100 mL^{-1} with four values out of five <40 UFC.100 mL^{-1} 1 mg.L^{-1} of residual chlorine after 30 min pH: 6.5–8.0
Sylviculture, turf **No consumable crop** **Restricted public access**	Secondary	–
Consumable crops in direct contact with reused water	Secondary + filtration, disinfection	Total coliforms <1,000 UFC.100 mL^{-1} with four values out of five <10,000 UFC.100 mL^{-1} pH: 5.6–8.0

Table 4.11. *Quality criteria of treated wastewater used for municipal irrigation in Australia*

4.8. Mediterranean countries and the Middle East

Recycling of wastewater for agricultural or other use is a common practice in most Mediterranean countries. For historical, economic and social reasons, the management of this practice differs greatly from one country to another and sometimes within the same country, such as in Italy or Spain.

Three groups of countries can be distinguished, each with its own guidelines:

1) no guidelines or regulations;

2) quality criteria based on those of the WHO;

3) standards similar to those of some US states and the US EPA.

The United Nations Environment Programme and the WHO issued guidelines for municipal wastewater reuse in the Mediterranean region in 2005 and 2006, respectively (United Nation Environment Programme 2005; WHO 2006).

These recommendations mainly address the microbiological risks associated with non-drinking uses in many countries. Five categories of uses have been created, and microbiological and physicochemical quality criteria, as well as treatments that can meet these criteria, are detailed. Criteria for nematodes and fecal coliforms or *E. coli* associated with irrigation use are summarized in Table 4.12.

Category[a]	**Use**[b]	**Nematodes (number of eggs L^{-1}, arithmetic mean)**	**Fecal coli or *E. coli* (UFC.100 mL^{-1}, geometric mean)**
I	Irrigation illimited access areas (golf courses, parks, sports fields)	≤0.1	≤200
III	Irrigation areas with no public access	≤1	≤105
IV	Underground irrigation category I areas	No recommendation	No recommendation

Table 4.12. *Recommendations for the microbiological quality of reused wastewater in the Mediterranean region (United Nation Environment Programme 2005; World Health Organization 2006)*

COMMENTS ON TABLE 4.12.–

[a]*Category II is not mentioned, as it includes uses other than irrigation.*

[b]*For the same use, recommendations vary depending on the irrigation method. It is a control point or risk minimization added to effluent quality criteria. Subsurface irrigation (category IV) is considered less at risk of exposure than spray irrigation (Category I).*

If poorly controlled, irrigation with treated wastewater can damage soils and lead to groundwater contamination.

The irrigation water quality is particularly important in arid areas with high evaporation rates due to the salt accumulation in the soil profile. Nutrients, such as nitrogen and phosphorus, are beneficial for plants and reduce the need for artificial fertilizers.

4.9. Conclusion

The growing scarcity of water in the world underlines the urgency to implement treated wastewater reuse concepts. Many countries with high water stress for decades have already responded and introduced appropriate regulations (Table 4.13) for water reuse to ensure sufficient water quality. Although there are wide differences between countries for the same irrigation use, the key is to believe in this concept and exploit all opportunities for treated wastewater reuse.

Each country has adapted reuse criteria according to its own model, which takes into account various local parameters such as the nature and quality of wastewater, physical and climatic conditions, sanitation capacities and different tertiary treatments for different uses and the associated costs.

In all cases, the health risk is the first parameter that led authorities to develop their criteria. For example, Algerian law prohibits the watering of all food crops intended to be consumed raw by humans.

Reuse of treated wastewater helps reduce the immediate deficit. However, the additional challenge is the implementation of treated wastewater reuse concepts. Only a few countries have implemented integrated water management concepts and planning systems.

Country	Type A example Food consumed raw			Type A example Watering of green spaces and sports fields with controlled public access			Type B example Cereal and fodder crops		
	BOD5 SS Turbidity	*E. coli.*/100 mL	*Legionella*/100 mL Nematodes L^{-1} Enterococci 100 mL^{-1}	BOD5 SS Turbidity	*E. coli* 100 mL^{-1}	*Legionella* 100 mL^{-1} Nematodes L^{-1} Enterococci 100 mL^{-1}	BOD5 SS Turbidity	*E. coli* 100 mL^{-1}	*Legionella* 100 mL^{-1} Nematodes L^{-1} Enterococci 100 mL^{-1}
WHO 2012	Turbidity <2	200	Nematodes $\leq 1\ L^{-1}$	SS: 20 $mg{\cdot}L^{-1}$ BOD5: 20 $mg{\cdot}L^{-1}$	1,000	Nematodes $\leq 1\ L^{-1}$	–	1,000	Nematodes $\leq 1\ L^{-1}$
US EPA 2012	BOD5 $\leq 10\ mg{\cdot}L^{-1}$ Turbidity ≤ 2 NTU	Undetectable	–	SS: 30 $mg{\cdot}L^{-1}$ BOD5: 20 $mg{\cdot}L^{-1}$	200	–	SS: 30 $mg{\cdot}L^{-1}$ BOD5: 20 $mg{\cdot}L^{-1}$	200	Requested in some states
EC 2020	Turbidity ≤ 5 NTU SS: 10 $mg{\cdot}L^{-1}$ BOD5: 10 $mg{\cdot}L^{-1}$	10	*Legionella*: 1,000 Nematodes $\leq 1\ L^{-1}$	SS: 35 $mg{\cdot}L^{-1}$ BOD5: 25 $mg{\cdot}L^{-1}$	100	*Legionella*: 1,000 Nematodes $\leq 1\ L^{-1}$	SS: 35 $mg{\cdot}L^{-1}$ BOD5: 25 $mg{\cdot}L^{-1}$	1,000	*Legionella*: 1,000 Nematodes $\leq 1\ L^{-1}$

Country	Type A example Food consumed raw			Type A example Watering of green spaces and sports fields with controlled public access			Type B example Cereal and fodder crops		
Middle East									
Jordan 2006	–	100	–	50	100	–	300	–	–
Israel	15 $mg \cdot L^{-1}$ SS	2.2–12	–	30 $mg \cdot L^{-1}$ SS	250	–	40 $mg \cdot L^{-1}$ SS	–	–
Syria	–	–	–	–	1,000	–	–	Non applicable	–
Saudi Arabia 2006	10 $mg \cdot L^{-1}$ SS	2.2	–	–	–	–	–	1,000	–
Kuwait	10 $mg \cdot L^{-1}$ SS	20	–	10 $mg \cdot L^{-1}$ SS	100	–	10 $mg \cdot L^{-1}$ SS	100	–
United Arab Emirates 2010	–	–	–	10 $mg \cdot L^{-1}$ SS	100	40	10 $mg \cdot L^{-1}$ SS	100	40
Egypt	–	Prohibited	–	5	100	–	30–60 $mg \cdot L^{-1}$ SS	1,000–5,000	–
Qatar 2005	–	–	–	5 $mg \cdot L^{-1}$ SS	100 in TC	–	5 $mg \cdot L^{-1}$ SS	100 in TC	–
Oman 1993	15 $mg \cdot L^{-1}$ SS	200	–	15 $mg \cdot L^{-1}$ SS	200	–	15 $mg \cdot L^{-1}$ SS	1,000	–

Country	Type A example Food consumed raw			Type A example Watering of green spaces and sports fields with controlled public access			Type B example Cereal and fodder crops		
				North Africa					
Morocco 2002	–	1,000[2]	–	100–200 mg·L^{-1}	200	–	–	1,000	–
Tunisia 2018	–	–	–	–	–	–	SS: 30 mg·L^{-1} BOD5: 20 mg·L^{-1}	–	Nematodes < 1 L^{-1}
Algeria 2012	Prohibited			10 mg·L^{-1} SS	200	–	15 mg·L^{-1} SS	1,000	–
				Europe					
France 2020	15 mg·L^{-1} BOD5 15 mg·L^{-1} SS	250	4 log + 6-log RNA phages[3] + 5-log SRA spores[3]	15 mg·L^{-1} SS	250	4 log + 4-log RNA phages + 4-log SRA spores	25 mg·L^{-1} SS	10,000	3 log + 3-log RNA phages + 3-log SRA spores
Spain 2006	Turbidity ≤ 10 NTU SS: 20 mg·L^{-1}	200	Nematodes, 1 u·L^{-1}	Turbidity ≤ 10 NTU SS: 20 mg·L^{-1}	200	Nematodes, 1 u·L^{-1}	SS: 35 mg·L^{-1}	1,000	Nematodes, 1 u·L^{-1}

Country	Type A example Food consumed raw			Type A example Watering of green spaces and sports fields with controlled public access			Type B example Cereal and fodder crops		
Greece	Turbidity ≤ 5 NTU SS: 10 mg·L^{-1} BOD5: 10 mg·L^{-1}	5	–	Turbidity ≤ 2 NTU SS: 10 mg·L^{-1} BOD5: 10 mg·L^{-1}	5	–	SS: 35 mg·L^{-1} BOD5: 25 mmg·L^{-1}	200	–
Asia									
China	Turbidity < 5 BOD5: 20 mg·L^{-1}	200	–	Turbidity < 5 BOD5: 20 mg·L^{-1}	200	–	SS: 90 mg·L^{-1} BOD5: 80 mg·L^{-1}	40,000	–
Japan	Turbidity ≤ 2	ND	–	Turbidity ≤ 2	ND	–	Turbidity ≤ 2	1,000	–

Table 4.13. *Reused treated wastewater quality in agriculture in various countries. [1]Depending on the irrigation method (with and without sprinkler). [2]Absence of salmonella, choleric vibrion, parasites and parasitic larvae. [3]On the whole wastewater treatment chain*

4.10. References

Abu-Rizaiza, O.S. (1999). Modification of the standards of wastewater reuse in Saudi Arabia. *Water Research*, 33(11), 2601–2608.

Ait Melloul, A. and Hassani, L. (1999). Salmonella infection in children from the waste-water spreading zone of Marrakech city (Morocco). *Journal of Applied Microbiology*, 87, 536–539.

Al-Jasser, A.O. (2011). Saudi wastewater reuse standards for agricultural irrigation: Riyadh treatment plants effluent compliance. *Journal of King Saud University, Engineering Sciences*, 23, 1–8.

Andreadakis, A., Gavalaki, E., Mamais, D., Tzimas, A. (2003). Wastewater reuse in Greece. *Global Nest Journal*, 5(1), 9–14.

Angelakis, A.N., Marecos Do Monte, M.H.F., Bontoux, L., Asano, T. (1999). The status of wastewater reuse practice in the Mediterranean basin: Need for guidelines. *Water Research*, 33(10), 2201–2217.

Armon, R. (2002). Surface and subsurface irrigation with effluents of different qualities and presence of Cryptosporidium oocysts in soil and on crops. *Water Science and Technology*, 46(3), 115–122.

Asano, T. and Levine, A.D. (1998). Wastewater reclamation, recycling, and reuse: An introduction. In *Wastewater Reclamation, Recycling and Reuse*, Asano, T. (ed.). CRC Press, New York.

Asano, T., Maeda, M., Tasaki, M. (1996). Wastewater reclamation and reuse in Japan: Overview and implementation examples. *Water Sci. Technol.*, 34, 219–226.

Australian Academy of Technological Sciences and Engineering (2004). Water Recycling in Australia. Report, Australian Academy of Technological Sciences and Engineering, Melbourne.

Ayres, R.M. (1992). Contamination of lettuces with nematode eggs by spray irrigation with treated and untreated wastewater. *Water Science and Technology*, 26(7–8), 1615–1623.

Ayers, R.S. and Westcot, D.W. (1985). *Water Quality for Agriculture*. FAO, Rome.

Barbagallo, S., Cirelli, G.L., Indelicato, S. (2001). Wastewater reuse in Italy. *Water Science and Technology*, 43(10), 43–50.

Biswas, A.K. and Arar, A. (1988). *Treatment and Reuse of Wastewater, Food & Agricultural Organization*. Butterworths, London.

Blumenthal, U.J. (2001). The risk of enteric infections associated with wastewater reuse: The effect of season and degree of storage of wastewater. *Transactions of the Royal Society of Tropical Medicine and Hygiene*, 95, 1–7.

Blumenthal, U.J., Mara, D.D., Peasey, A., Ruiz-Palacios, G., Stott, R. (2000). Guidelines for the microbiological quality of treated wastewater used in agriculture: Recommendations for revising WHO guidelines: Environment and health. *Bulletin of the World Health Organization*, 78(9), 1104–1116.

Bouwer, H. (1991). Groundwater recharge with sewage effluent. *Water Science and Technology*, 23, 2099–2108.

Bridgeman, J. (2004). Public perception towards water recycling in California. *The Journal*, 18(3), 150–154.

California Code of Regulations (2018). State Water Resources Control Board Regulations Related to Recycled Water. Report, California Code of Regulations, California.

California State (2009). Treatment technology report for recycled water. Department of Public Health, Division of Drinking Water and Environmental Management, Recycled Water Unit, Carpinteria.

California State Water Resources Control Board (2020). Municipal wastewater recycling survey. California State Water Board, Sacramento.

Camann, D.E. and Moore, B.E. (1987). Viral infections based on clinical sampling at a spray irrigation site. In *Implementing Water Reuse. Proceedings of Water Reuse Symposium IV*. American Water Works Association, Denver.

Chang, A.C., Page, A.L., Asano, T. (1995). Developing human health-related chemical guidelines for reclaimed wastewater and sewage sludge applications in agriculture. Report, World Health Organization, Geneva.

Cifuentes, E. (1998). The epidemiology of enteric infections in agricultural communities exposed to wastewater irrigation: Perspectives for risk control. *International Journal of Environmental Health Research*, 8, 203–213.

Cooper, R.C., Olivieri, A.W., Cahn, M.D., Colford, J., Crook, J., Debroux, J.F., Mandrell, R., Suslow, T., Tchobanoglous, G. (2012). Review of California's water recycling criteria for agricultural irrigation. Final Report, National Water Research Institute (NWRI).

Cordy, G. (2003). Persistence of pharmaceuticals, pathogens, and other organic waste-water contaminants when wastewater is used for groundwater recharge. In *Proceedings of the Third International Conference on Pharmaceuticals and Endocrine Disrupting Chemicals in Water*, Minneapolis.

Crook, J. (1998). Water reclamation and reuse criteria. In *Wastewater Reclamation, Recycling and Reuse*, Asano, T. (ed.). US EPA, New York.

Crook, J. and Surampalli, R.Y. (2005). Water reuse criteria in the United States. *Water Science and Technology: Water Supply*, 5(3/4), 1–7.

Crook, J., MacDonald, J.A., Trussell, R.R. (1999). Drinking use of reclaimed water. *Journal of the American Water Work Association*, 91(8), 40–49.

Directives Européennes (2020). Règlement (UE) no. 2020/741 du 25/05/20 relatif aux exigences minimales applicables à la réutilisation de l'eau. *JOUE*, 177.

Drewes, J.E., Heberer, T., Reddersen, K. (2002). Fate of pharmaceuticals during indirect drinking reuse. *Water Science and Technology*, 46(3), 73–80.

EU Water Directors (2016). Common implementation strategy for the water framework directive and the floods directive. Guidelines on integrating water reuse into water planning and management in the context of the WFD. Report, European Environment Agency, Copenhagen.

Exall, K., Marsalek, J., Schaefer, K. (2004). A review of water reuse and recycling, with reference to Canadian practice and potential: Incentives and implementation. *Water Quality Research Journal*, 39(1), 1–12.

Fattal, B., Lampert, Y., Shuval, H. (eds) (2004). A fresh look at microbial guidelines for wastewater irrigation in agriculture: A risk-assessment and cost-effectiveness approach. In *Wastewater Use in Irrigated Agriculture*. CAB, Wallingford.

Florida Department of Environmental Protection (FDEP) (2021). Reuse of reclaimed water and land application. Report, Florida Administrative.

Gauvin, D. (2006). Avis sur la réutilisation d'effluents de stations de traitement d'eaux usées domestiques pour l'irrigation d'un terrain de golf. Report, Institut national de santé publique du Québec, Quebec.

Gerba, C.P. and Rose, J.B. (2003). International guidelines for water recycling microbiological considerations. *Water Science and Technology: Water Supply*, 3(4), 311–316.

Haas, C.N. (1983). Estimation of risk due to low doses of microorganisms: A comparison of alternative methodologies. *American Journal of Epidemiology*, 118(4), 573–582.

Haas, C.N., Rose, J.B., Gerba, C.P. (1999). *Quantitative Microbial Risk Assessment*. John Wiley & Sons, New York.

Huot, C., Gauvin, D., Lévesque, B., Payment, P. (2008). Portrait Québécois et évaluation du risque à la santé de la réutilisation des eaux usées domestiques traitées pour l'irrigation de grandes surfaces gazonnées. Report, Institut National de la santé publique du Québec, Quebec.

Jaramillo, M.F. and Restrepo, I. (2017). Wastewater reuse in agriculture: A review about its limitations and benefits. *Sustainability*, 9(10), 1734–1745.

Jodar-Abellan, A., Lopez-Ortiz, M.I., Melgarejo-Moreno, J. (2019). Waste water treatment and reuse in Spain. *Water*, 11, 1551–1162.

Lazarova, V., Levine, B., Sack, J., Cirelli, G., Jeffrey, P., Muntau, H., Salgot, M., Brissaud, F. (2001). Role of water reuse for enhancing integrated water management in Europe and Mediterranean countries. *Water Science and Technology*, 43(10), 25–33.

Lazarova, V., Gaid, A., Rodriguez-Gonzales, J., Alday Ansola, J. (2003). L'intérêt de la réutilisation des eaux usées : analyses d'exemples mondiaux. *Techniques, sciences et méthodes*, 9, 64–85.

Lee, H. and Tan, T.P. (2016). Singapore's experience with reclaimed water: NEWater. *International Journal of Water Resources Development*, 32(4), 611–621.

Lyu, S., Chen, W., Zhang, W., Fan, Y., Jiao, W. (2016). Wastewater reclamation and reuse in China: Opportunities and challenges. *Journal of Environmental Science*, 39, 86–96.

Mediterranean Action Plan (2005). Guidelines for municipal water reuse in the Mediterranean region. Report, United Nations Environment Program in collaboration with the World Health Organization, Athens.

Ministère de l'écologie, de l'énergie, du développement durable et de la mer (2010). Arrêté du 2 août 2010 relatif à l'utilisation d'eaux issues du traitement d'épuration des eaux résiduaires urbaines pour l'irrigation de cultures ou d'espaces verts. Report, Ministère de l'écologie, de l'énergie, du développement durable et de la mer.

Ministère de la transition écologique (2022). Décret no 2022-336 du 10 mars 2022, relatif aux usages et aux conditions de réutilisation des eaux usées traitées. Ministère de la transition écologique.

Neubert, S. and Benabdallah, S. (2003). La réutilisation des eaux usées traitées en Tunisie. Studies and expert reports, German Institute of Development and Sustainability, Bonn.

Nurizzo, C. (2003). Reclaimed water reuse in the Mediterranean region: Some considerations on water resources, standards and bacterial re-growth phenomena. *Water Science and Technology: Water Supply*, 3(4), 317–324.

Odendaal, P.E., van der Westhuizen, J.L.J., Grobler, G.J. (1998). Wastewater reuse in South Africa. In *Wastewater Reclamation, Recycling and Reuse*, Asano, T. (ed.). CRC Press, New York.

Ogoshi, M., Suzuki, Y., Asano, T. (2001). Water reuse in Japan. *Water Science and Technology*, 43(10), 17–23.

Payment, P. and Pintar, K. (2006). Waterborne pathogens: A critical assessment of methods, results and data analysis. *Revue des sciences de l'eau*, 19(3), 233–245.

Petterson, S.A. and Ashbolt, N.J. (2003). WHO Guidelines for the safe use of wastewater and excreta in agriculture: Microbial risk assessment section. Report, World Health Organization, Geneva.

Radcliffe, J.C. and Page, D. (2020). Water reuse and recycling in Australia – history, current situation and future perspectives. *Water Cycle*, 1, 19–40.

Rotbardt, A. (2011). REUT–Perspectives opérationnelles et recommandations pour l'action. Report, Agence Française de Développement and BRL Ingénierie.

Schwartzbrod, J. (1989). Impact of wastewater treatment on helminth eggs. *Water Science and Technology*, 21(3), 295–297.

Sheikh, B., Cooper, R.C., Israel, K.E. (1999). Hygienic evaluation of reclaimed water used to irrigate food crops – A case study. *Water Science and Technology*, 40(4–5), 261–267.

Shoushtarian, F. and Negahban-Azar, M. (2020). Global regulations and guidelines for agricultural water reuse: A critical review. *Water*, 12, 971–985.

Takeuchi, H. and Tanaka, H. (2020). Water reuse and recycling in Japan – History, current situation, and futures perspectives. *Water Cycle*, 1, 1–11.

Tortajada, C. and van Rensburg, P. (2020). Drink more recycled wastewater. *Nature*, 577(7788), 26–28.

United Nation Environment Programme (2005). Guidelines for municipal water reuse in the Mediterranean region Mediterranean action plan. Report, United Nation Environment Programme, Athens.

U.S. Environmental Protection Agency, U.S. Agency for International Development (2004). Guidelines for Water Reuse. Report, U.S. Environmental Protection Agency et U.S. Agency for International Development, Washington.

U.S. Environmental Protection Agency, U.S. Agency for International Development (2012). Guidelines for Water Reuse. Report, U.S. Environmental Protection Agency and U.S. Agency for International Development, Washington.

Washington State Department of Ecology (2018). Use of recycled water for the irrigation of crops. Report, Administrative Code 173-219-390.

Water Unit Environmental Health Directorate (2011). Guidelines for the non-drinking uses of recycled water in Western Australia. Report, Water Unit Environmental Health Directorate.

Westrell, T., Schönning, C., Stenström, T.A., Ashbolt, N.J. (2004). QMRA (quantitative microbial risk assessment) and HACCP (hazard analysis and critical control points) for management of pathogens in wastewater and sewage sludge treatment and reuse. *Water Science and Technology*, 50(2), 23–30.

WHO (1989). Health guidelines for the use of wastewater in agriculture and aquaculture. Technical Report no. 778, WHO, Geneva.

WHO (2006a). *WHO Guidelines for the Safe Use of Wastewater, Excreta and Greywater. Volume 2: Wastewater Use in Agriculture*. WHO, Geneva.

WHO (2006b). A compendium of standards for wastewater reuse in the Eastern Mediterranean Region. Regional Office for the Eastern Mediterranean. Report, WHO, Cairo.

WHO (2012). *WHO Guidelines for the Safe Use of Wastewater, Excreta and Greywater. Volume 2: Wastewater Use in Agriculture*. WHO, Geneva.

Yates, M.V. and Gerba, C.P. (1998). Microbial considerations in wastewater reclamation and reuse. In *Wastewater Reclamation, Recycling and Reuse*, Asano, T. (ed.). CRC Press, New York.

5

Treated Wastewater Reuse Planning

Treated wastewater reuse projects do not only include technical aspects, but may have environmental, economic, social implications and impacts on public health. Planning for these projects should incorporate a number of areas such as market assessment and public acceptance.

Due to the complexity and sensitivity of treated wastewater reuse, this process should take place in a number of stages from technical design to construction:

– scope, objectives and delimitation of the area including the project;

– preliminary investigations;

– assessment of potential markets;

– assessment of alternatives;

– user and institution agreements;

– design;

– construction;

– exploitation and evaluation.

The first four steps define the project feasibility. The following five steps cover actual project implementation.

5.1. Project objectives and limitations

Reuse projects can be integrated into master plans. Each project should begin with a definition of its scope and objectives. In principle, a water reuse project is versatile, as it is at least both a clean-up project and a water supply project (drinking or non-drinking). In all cases, public health will be the dominant issue, which will have to be ensured by protecting users.

5.1.1. *Preliminary investigations*

This step is dedicated to the collection of information and lists all potential wastewater sources that can be reused and all potential applications for the project.

Generally, the following should be addressed:

– local sources of (treated) wastewater suitable for reuse;

– potential local uses for treated wastewater;

– current uses of freshwater resources;

– current and future costs of reused treated wastewater and fresh water;

– public health considerations associated with reuse of treated wastewater;

– potential environmental impacts of reuse of treated wastewater;

– laws, regulations and legal obligations affecting the reuse option;

– identification of local and national institutions involved in regulations;

– sources of funding available to support the reuse program;

– general survey on public attitudes, acceptance or opposition, as well as religious concerns.

All this information should be based on a solid project in which the economic, health and social aspects are well defined.

5.1.2. *Selection of potential markets*

When freshwater volumes are difficult to sustain for specific reasons (drought and water stress) and they are allocated to priority destinations, an economic comparison between the cost of producing freshwater and the cost of reusing treated wastewater should be made:

– estimate of water quality and quantity required for each type of application;

– verification of regulatory requirements for each type of application;

– estimate of the quality of the reused treated wastewater (based on objectives) and comparison with regulatory requirements and users' requirements;

– estimate of the amount of reusable water and ability to respond to changing demands for each type of application;

– estimated cost of treatment and distribution of reused treated wastewater for each type of application.

In addition to the comparison of costs, other criteria derived from technical, social and environmental data can be used in the selection process (e.g. desalination, institutional complexity, reliability, energy consumption and user acceptance).

5.1.3. *Assessment of alternatives*

The most likely alternatives are assessed in more detail, including a detailed study of potential users, technical optimization, construction and operating costs.

The following information should be obtained:

– precise uses of the reused of treated wastewater;

– current and future quantitative needs;

– quality needs and specific use constraints;

– modifications to on-site facilities required to obtain wastewater compliant with expected standards and an estimate of the associated costs.

Technical aspects should include system reliability and application of multiple barriers in the elimination of substances.

The public information program starts as soon as the project is implemented with the first educational activities targeted for the general public. Also, the environmental protection process will be developed and/or an assessment of the impact on public health will be estimated.

Public awareness and education programs are a particularly important aspect of water reuse projects (Table 4.1), as there is a close association in the public mind between water quality and health, while there is a general lack of knowledge about the wastewater treatment process.

5.1.4. *Awareness and education*

It is about meeting the direct and natural needs of the public. Transparency should inspire public trust and improve perception and acceptance: this is essential to overcome social and cultural barriers.

Meetings, workshops and conferences are means of informing and reassuring the public. Once the project is implemented and completed, the aim of the educational activities is to influence users to make full use of the volumes of water produced.

Any early public participation will provide insight into future attitudes toward a treated wastewater water reuse project.

A citizens' advisory committee can help to ensure the technical aspects of a project.

Local or regional media coverage can be important for a project. As such, information should be provided to the media.

Type of use for reused treated wastewater water	% of people opposed to this use
Drinking water	55
Food preparation	50
Showers and baths	35
Swimming pools, lakes	21
Laundry	21
Irrigation of cereals, vineyards, etc.	12
Flushing toilets	5
Irrigation of parks, golfs, etc.	4

Table 5.1. *Percentage of people opposed to reuse of treated wastewater in the US (Davies and Hirji 2003)*

The public perception of agricultural products irrigated with treated wastewater is generally negative based mostly on hygiene and health-related themes, the lack of confidence in public quality control, as well as a natural aversion to these foods that have been potentially in contact with human excreta. The consequence is that farmers are unable to sell their crop, even if it meets the important quality criteria required, making them reluctant to use treated wastewater for the irrigation of their crops. Public awareness and trust are prerequisites for successful reuse of treated wastewater, so information sessions on all aspects (important environmental and economic benefits, precautions to be taken to make the reuse of treated wastewater safe and reliable, etc.) would be necessary to reinforce social acceptance of treated wastewater reuse.

5.1.5. *Importance of terminology*

It is important to use the terminology of water reuse carefully with the public. For example, "water" and "wastewater" suggest radically different properties and are perceived differently by the general public.

The fact that reused water is produced from wastewater should not be concealed, and information should carefully distinguish between "reuse of wastewater" and "reuse of treated wastewater".

The omission of this last term could give the impression that raw wastewater will be used. In fact, the treated wastewater will be subject to additional treatments with advanced technologies that will produce a quality water similar to freshwater.

Risk control is an important factor. People are willing to take more risk if they perceive that operators manage quality control well and all distributed water analyses are available at all times.

5.1.6. *Economic and legal aspects*

The treated wastewater reuse projects are diverse and may involve sanitation improvements, such as wastewater collection, water pollution control, such as the reduction of waste discharges in watercourses, alternative water production, and social or economic improvements, such as the development of an agricultural area.

The benefits included in the economic analysis depend on the objectives of the project. Advantages may include environmental benefits, such as reduction of oxygen depletion in watercourses, social/economic benefits, production increase and a reduction of fertilizer costs.

Reuse of treated wastewater may not be perceived by government as a priority if short-term direct benefits are not evident.

To avoid the costs of the treated wastewater reuse alternative, the following estimates should be included in the alternative's assessment:

– additional treatment of wastewater at the treatment plant outlet, according to the standards assigned to the different applications;

– collection and transport of treated wastewater ready for reuse to points of use;

– operating cost;

– waste fate.

Additional treatments are minimal if effluents are reused, especially for agriculture where nutrients are beneficial.

If the final treated wastewater is intended for drinking, advanced and particularly reliable treatment should be implemented (see Chapter 6).

5.2. References

Chang, A.C., Page, A.L., Asano, T., Hespanhol, I. (1996). Developing human health-related chemical guidelines for reclaimed wastewater irrigation. *Water Science & Technology*, 33(10/11), 463–472.

Davies, R. and Hirji, R. (2003). Water resources and environment; technical note: F3, wastewater reuse. World Bank, Washington.

Khouri, N., Kalbermatten, J.M., Bartone, C.R. (1994). The reuse of wastewater in agriculture: A guide for planners. Report, Water and Sanitation Report No. 6, UNDP-World Bank Water and Sanitation Program. World Bank, Washington.

Metcalf & Eddy Inc. (ed.). (1991). *Wastewater Engineering Treatment, Disposal, and Reuse.* McGraw-Hill, New York.

Rosenblum, E. (1999). Selection and implementation of non potable water recycling in "Silicon Valley" (San Jose area) California. *Water Science & Technology*, 40(4–5), 51–57.

Rowe, D.R. and Abdel-Magid, I.M. (1995). *Handbook of Wastewater Reclamation and Reuse*. Lewis Publishers, Boca Raton.

U.S. Environmental Protection Agency, Agency for International Development (1992). Guidelines for water reuse. Report, U.S. Environmental Protection Agency and Agency for International Development, Washington.

World Health Organization (1986). Guidelines for planning community participation activities in water supply and sanitation projects. Report, World Health Organization, Geneva.

6

Treatment Technologies

Many tertiary treatment options are used around the world, and each process has its advantages but also its limitations or disadvantages. The choice of one process over another takes into account technical, economic aspects.

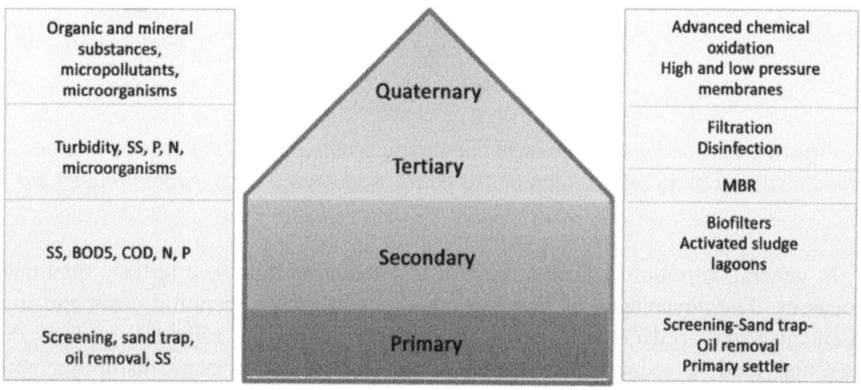

Figure 6.1. *Performance distribution depending on specific treatments*

Tertiary treatment is necessary for irrigation and non-drinking urban uses. It allows to reach levels of physical and bacteriological quality that meet the standards governing treated wastewater reuse, especially in agriculture, industry (excluding food) and aquatic ecosystems (water bodies, wetlands, estuaries, slow-flowing rivers, etc.).

Quaternary treatment is essential for drinking water uses (direct, groundwater recharge, food industry, etc.). It provides peace of mind knowing that treated wastewater is clean, hygienic, compliant with regulations and safe enough to be consumed or reused.

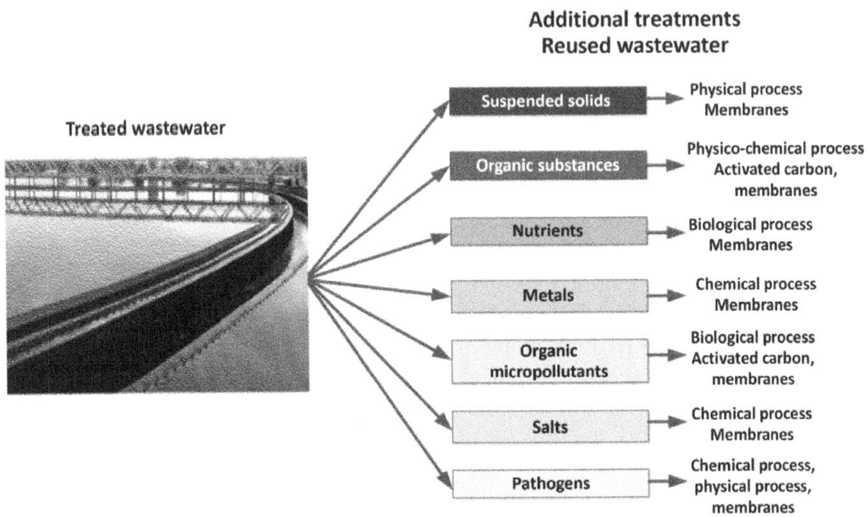

Figure 6.2. *Additional treatments depending on the uses of the reused treated wastewater. For a color version of this figure, see www.iste.co.uk/gaid/reuse1.zip*

In general, filtration, chlorination and UV radiation dominate tertiary treatment processes. The advantages of these technologies are their operating costs and low greenhouse gas (GHG) emissions. Ozonation is highly efficient against the microbial load to reduce pharmaceutical residues and endocrine disruptors with moderate processing costs. Peracetic acids are gradually implemented.

A variety of different technologies are available to achieve optimal effluent quality for various uses. These processes involving the use of membranes are described mainly in terms of their ability to remove pathogens and dissolved mineral and organic substances.

The quaternary treatment adds a fourth level of treatment, more advanced and rigorous. For example, using high-pressure membranes, it leads to water purification levels for reuse in drinking water manufacturing processes.

In all cases, when choosing the treatment process, the parameters considered are quality objectives, economic feasibility, environmental impact, social perspective and local customs.

6.1. Use of raw wastewater

In many countries, wastewater for crop irrigation is most often used without treatment and it is also common that no other health protection measure is provided for to limit as much as possible negative impacts on the health of workers and consumers.

The raw wastewater used in agriculture actually causes various health problems, particularly helminthiasis and diarrheal diseases in children and adults.

Direct use of untreated wastewater occurs in areas where treatment plants are not in place and alternatives are limited, usually in arid or semi-arid areas. For example, these practices may involve direct pumping of raw wastewater into the pipe before it arrives at the treatment plant. Although officially prohibited in most countries, direct use of untreated wastewater occurs in many urban and peri-urban areas.

Only drip irrigation systems can reduce negative impacts. If raw wastewater is used for unrestricted irrigation, the cultivated products must be disinfected and can only be consumed after cooking but it is a particular case forbidden in most of countries. Disinfection efficiency is still very limited despite high dosages in chlorine. The real solution is to change behaviors and habits.

Vaccination campaigns (especially against typhoid, polio and hepatitis A) are effective measures to prevent and reduce morbidity, but they cannot be considered as alternatives to wastewater treatment and other health protection measures. They are intended to supplement the health protection measures adopted. The first solution is to prohibit such practices, since it is difficult to be protected despite the various health protection measures that can be applied.

Restrictions regarding crops must be high and worker exposure must be limited.

Whatever disinfection system is considered, it will not be effective due to the very high concentration of suspended matter and organic substances, which will induce too high of an immediate oxidant demand.

6.2. Bases of the treatment process concept

Water reuse generally refers to the process of using treated wastewater for beneficial purposes such as agricultural and landscape irrigation, industrial processes, non-drinking urban applications (such as toilet water, street washing and fire protection), groundwater recharge, leisure and direct or indirect water supply.

Conventional municipal wastewater treatment consists of a combination of physical, chemical and biological processes and operations to remove solids, organic matter, pathogens, toxic metals and sometimes nutrients from wastewater. The different treatment steps include preliminary (coarse), primary and secondary treatments.

Tertiary treatment of wastewater works by using a combination of physical and chemical processes to remove suspended particles and microbiological agents. The process usually involves filtration followed by a disinfection treatment. In some cases, tertiary treatment may also use other specialized treatments such as lagoon storage, elimination of biological nutrients and elimination of nitrogen and phosphorus. The elimination of these substances makes water ready for reuse.

The technologies are largely derived from those used for conventional wastewater treatment and for the manufacture of drinking water.

The objective of the design of a technical branch for treated wastewater reuse is to develop an integrated treatment system capable of reliably meeting water quality objectives.

To achieve this objective, the characterization of water quality is necessary to assess the biological and chemical safety of the use of reused treated wastewater for various uses, as well as to determine the reliability of the implemented technologies.

The water quality parameters used to assess reused treated wastewater are relevant parameters for monitoring water quality (Table 6.1). Municipal wastewater treatment is designed to meet quality objectives on biochemical oxygen demand (BOD5), suspended solids (SS), total or fecal coliforms and nutrient levels (nitrogen and phosphorus).

In the reuse principle, monitoring programs include microorganisms, such as bacteria, viruses, protozoa, turbidity, dissolved minerals, disinfection sub-products and specific inorganic and organic contaminants.

Parameter	Importance in treated wastewater reuse	Approximative range in treated wastewater	Objective of reused wastewater treatment
Organic indicators			
BOD5	Organic substrate	10–30 $mg \cdot L^{-1}$	<1–10 $mg \cdot L^{-1}$
TOC	Total organic carbon	1–20 $mg \cdot L^{-1}$	<1–10 $mg \cdot L^{-1}$
Particle measurements			
Total suspended solids (SS)	Reduces disinfection effectiveness	<1–30 $mg \cdot L^{-1}$	<1–10 $mg \cdot L^{-1}$
Turbidity	May be correlated to SS	1–30 NTU	0.1–10 NTU
Microorganisms			
Pathogenic organisms	Major health risks	Coliform organisms: 10^3–10^6 100 mL^{-1}	0–2,000 mL^{-1}
Nutriments			
Nitrogen	Nutrient source for irrigation	10–30 $mg \cdot L^{-1}$	<1–30 $mg \cdot L^{-1}$
Phosphorus	Nutrient source for irrigation	0.1–30 $mg \cdot L^{-1}$	<1–20 $mg \cdot L^{-1}$

Table 6.1. *Summary of the main parameters used to characterize reused treated wastewater quality*

The passage of raw wastewater through a biological treatment plant effectively eliminates most biodegradable matter, nutrients and pathogens, so that the treated wastewater then has a wide range of potential applications.

For a bacterial concentration in raw water of 10^6–10^8 u·100 mL^{-1}, the outlet concentration is between 10^5 and 10^7 u·100 mL^{-1}.

Treatment plants are mainly designed for the reduction of BOD5, suspended solids and often nutrients (nitrogen and phosphorus). They can also reduce some of the pathogens (bacteria, viruses and protozoa) by about 90% and helminth eggs by about two logarithmic units, according to the suspended solid concentration at the plant outlet.

To treat municipal wastewater for safe use, it is important to obtain greater reductions of pathogens.

Treatment process	Elimination of pathogens (log unit)			
	Viruses	Bacteria	Protozoa cysts or oocysts	Helminth eggs
Primary settler	0–1	0–1	0–1	0–1
Primary settler with coagulant	1–2	1–2	1–2	1–2
Activated sludge + sedimentation	1–2	1–2	0–1	1–1.5
Biological filter + sedimentation	1–2	1–2	0–1	1–2
Biological filter (fine material)	1–2	1–2	0–1	1–2.5
Membrane bioreactor (MBR)	1–2	≥5	≥5	6
Aerated lagoon + settler	1–2	1–2	0–1	2–3

Table 6.2. *Pathogen elimination efficiency at the treatment plant*

The tertiary treatment refers to treatment processes downstream of secondary treatment such as additional elimination of solids by coagulation, flocculation and sedimentation, granular media filtration and disinfection. Unfortunately, the tertiary treatment, and, even in some cases the secondary treatment, is sometimes unaffordable and often too complex to be satisfactorily used in many low-income countries.

Effective pathogen elimination requires a good choice of treatment processes, since several groups of pathogens – viruses, bacteria, protozoa and helminths – must be eliminated at different levels at the lowest possible cost.

These disinfection methods eliminate bacteria and viruses, but not helminth eggs as they are very resistant and behave differently from bacteria and viruses during treatment. Protozoan (oo)cysts are slightly less resistant than helminth eggs. Therefore, the selection process that will remove helminth eggs and protozoan (oo)cysts from wastewater to the required level must be carefully selected.

6.3. References

Bahri, A. and Brissaud, F. (2004). Setting up microbiological water reuse guidelines for the Mediterranean. *Water Science and Technology*, 50(2), 39–46.

Bixioa, D., Thoeyea, C., De Koningb, J., Joksimovicb, D., Savicc, D., Wintgensd, T., Melind, T. (2006). Wastewater reuse in Europe. *Desalination*, 187, 89–101.

Bourke, M. (2002). Disinfection of treated wastewater – Guidelines for environmental management. *EPA Victoria*, 370.

Crini, G. and Lichfouse, E. (2018). *Wastewater Treatment: An Overview. Green Adsorbents for Pollutant Removal*. Springer, Berlin.

Del Pino, M.P. and Durham, B. (2001). Wastewater reuse through dual membrane processes: Opportunities for sustainable water resources. *Desalination*, 124, 271–277.

Environmental Protection Agency (2012). Guidelines for Water Reuse. Report, Environmental Protection Agency.

Gaid, A. (1984). *Epuration biologique des eaux usées 1 et 2*. Office des publications universitaires, Alger.

López-Pila, J. and Szewzyk, R. (2000). Estimating the infection risk in recreational waters from the faecal indicator concentration and from the ratio between pathogens and indicators. *Water Research*, 34(17), 4195–4200.

Pangarkar, B.L., Parjane, S.B., Sane, M.G. (2010). Design and economical performance of gray water treatment plant in rural region. *World Academy of Science, Engineering and Technology*, 4, 765–769.

Salgota, M., Huertasa, E., Weberb, S., Dottb, W., Hollender, J. (2006). Wastewater reuse and risk: Definition of key objectives. *Desalination*, 187, 29–40.

Sirvain, M.A. (2012). Réutilisation de l'eau : place et enjeux des séparations liquides-solides et membranaires. *L'Eau, L'Industries, Les Nuisances*, 363, 87–91.

Thomazeau, R. (1981). *Stations d'épuration : eaux potables, eau usées : précis théoriques et technologique*. Technique et Documentation, Paris.

Wintgens, T., Melin, T., Schäfer, A., Khan, S., Muston, M., Bixio, D., Thoeye, C. (2004). The role of membrane processes in municipal wastewater reclamation and reuse. *Desalination*, 178(1–3), 1–11.

7

Tertiary Filtration

Regarding wastewater, tertiary filtration is a process used to eliminate suspended particles and particulate organic matter (particulate BOD) from the water at the outlet of the treatment plant (effluent) after sedimentation.

Filtration on granular media can be carried out by gravity (fast gravity filtration) under pressure (pressure filtration) or mechanical filtration (filter disks).

The filtration system is implemented:

– directly on the outlet treated wastewater;

– after coagulation flocculation on the outlet treated wastewater.

Filters are either made of a single layer of media (single-layer filter) or two media (dual-media filter).

The concentration of suspended solids (SS) on the treated wastewater varies on average between 5 and 35 $mg \cdot L^{-1}$. Filtration on filter media is then used as a refining step and delivers SS $\leq$ 5 $mg \cdot L^{-1}$ on the final wastewater and even less, depending on the filter and media design. Filtration systems should be operated in such a way as to reduce SS levels as much as possible, which can be expressed as turbidity (NTU). Filtration is therefore an important barrier to the removal of particles and some microorganisms. To ensure the most effective inactivation of pathogens during disinfection, the filtered water must be low in turbidity. Filtration systems must be designed so that their reliability ensures continuous logarithmic pathogen removal credits by disinfection following this step.

Treated wastewater can be reused for several purposes: irrigation, urban non-drinking uses, recreation areas, wetlands and possibly groundwater recharge.

Granular media filtration eliminates SS and colloidal particles. These include biochemical oxygen demand (particulate BOD5), chemical oxygen demand (particulate COD), bacteria and particulate chemical contaminants.

Filtration is a useful treatment step to eliminate protozoan (oo)cysts and helminth eggs from effluents. During filtration, pathogens and other particulate matter are partially eliminated when passing through the granular and porous material(s).

SS elimination is a recommended step to then ensure proper disinfection. It is particularly popular in ecologically sensitive areas or in irrigation, where regulatory requirements for discharges are strict.

7.1. Gravity filtration without reagents

The only physical treatment barrier is based on material characteristics and filter operating conditions.

This type of filtration, used in most cases, is a simple and economically interesting process.

The parameters used for filtration design are filtration rate, height and type of media, pressure drop and retention capacity.

In each case, the media height is to be considered in order to avoid that too high filtration rates quickly break through the filter due to a too low media height.

Rapid two-layer filtration can achieve 2-log *Giardia* cyst and *Cryptosporidium* oocyst removal credits provided the turbidity of filtered water is consistently below 1 NTU and 80% of the time at 0.5 NTU.

7.1.1. *Rapid filters and high-rate filters (HRF)*

The main filtration mechanism of rapid filters is surface retention, that is in the first third of the media height. The shape of the filter media also affects the retention capacity in that angular particles are preferable to rounded particles. Regardless of the choice of the support media, the size of the media is between 0.85 and 1.00 mm (effective size [ES]) for the most common cases. Figure 7.1 shows the design of a sand filter with the filter media support floor, media height, water production circuit and backwashing circuit.

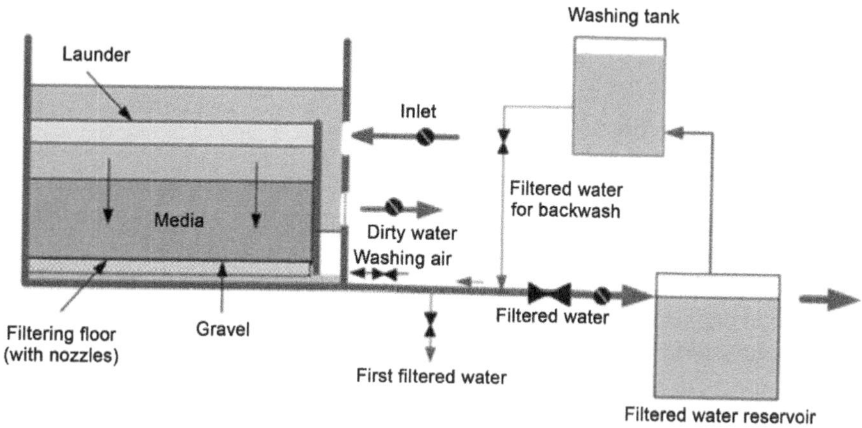

Figure 7.1. *Filter operating principle. For a color version of this figure, see www.iste.co.uk/gaid/reuse1.zip*

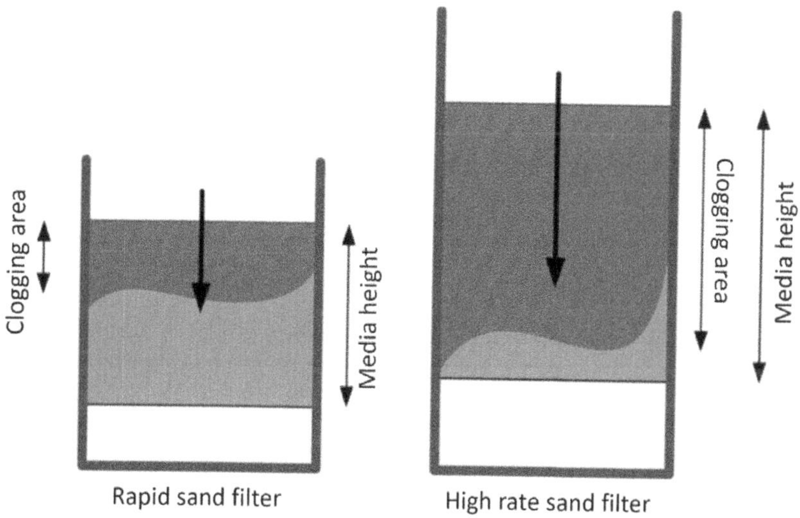

Figure 7.2. *Clogging depth between rapid sand filter and high-rate sand filter. For a color version of this figure, see www.iste.co.uk/gaid/reuse1.zip*

Over the last two decades, HRF have been developed. Their advantage is that they operate at high rate (HR; >15 m·h^{-1}) and have an important media height that allows for deep particle diffusion, ensuring greater retention capacity and longer cycle times (Figure 7.2).

7.1.2. *Dual-media filters*

Dual-media filters are made up of two media of different density and grain size, filtering in depth and operating at higher rates than conventional rapid sand filters (Figure 7.3). The gravel layer is not taken into account, as it does not act as a filter barrier because of its too high grain size (5–7 mm). However, it is necessary, on the one hand, to protect nozzles and, on the other hand, to ensure the homogenization of air and water when backwashing the filter.

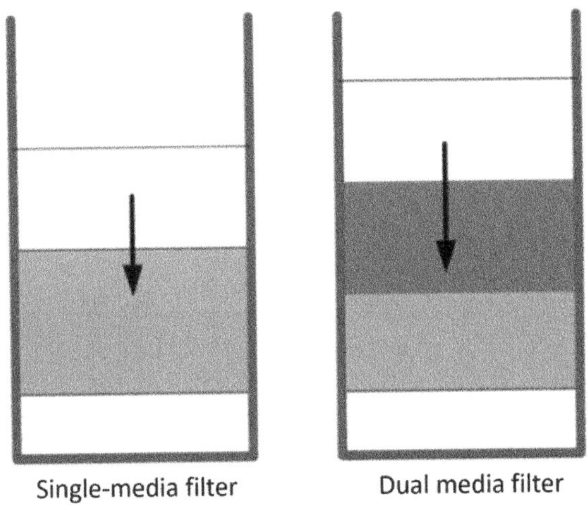

Figure 7.3. *Diagram of the different types of filters: single media and dual media. For a color version of this figure, see www.iste.co.uk/gaid/reuse1.zip*

For wastewater, dual-media tertiary filters can be made up of anthracite or pumice combined with sand. In these types of filters, the lower density of anthracite or pumice allows a larger grain size to be used so that after backwashing, the larger size anthracite or pumice is on top of the smaller effective-sized sand. In this way, the water is first filtered through the media with the larger ES and then reaches the media with the smaller ES, thus making better use of the total filtration height. Dual-media filters have longer cycle times, as their retention capacity is higher due to a large retention area.

When the filter is clean, particles are retained by transport mechanisms. During the filter fouling period, the particles already retained serve as a collector for the other particles, which increases the removal rate. Subsequently, this rate stabilizes for a certain period of time to finally deteriorate by causing the filter to be pierced due to an increase in turbidity at the output of the filter.

These filters can operate at filtration rates of up to 20 $m \cdot h^{-1}$, above which pressure losses develop too rapidly. The result is too short cycle times.

Dual-media filters have a very low cut-off threshold that can consistently eliminate 85%–90% of particles. This is due to the maturation of the upper filter layer, which can retain small particles titrant in turbidity, as well as due to an ES of the sand (lower layer), which is often smaller than that of a conventional sand filter.

Unlike the single-media filter with low ES sand, where the clogging occurs in a top layer of about 30–40 cm (resulting in an exponential increase in pressure loss), the dual-media filter allows for optimal use of the two layers.

Because of its grain size, the top layer of anthracite works according to the principle of volume filtration, that is to say it retains particles over its entire depth.

The height of large media (top layer) acts as a reservoir and storage for retained particles. The sand layer (bottom layer) ensures the filtration finish by retaining small particles. It also helps to ensure protection against the risks of breakthrough.

Due to the deep particle retention, the pressure drop of a dual-media filter increases more slowly than that of a sand filter.

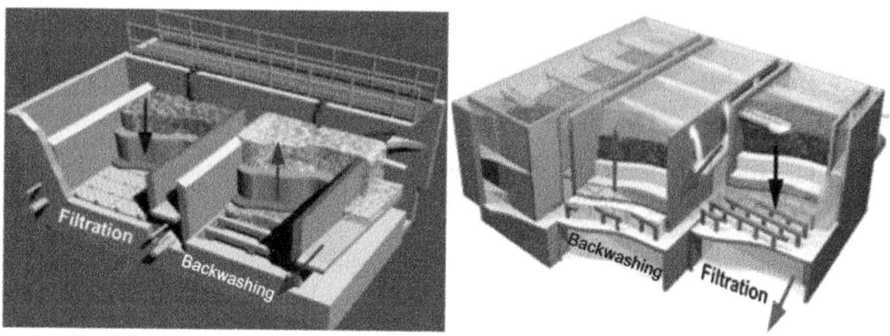

Figure 7.4. *Single-media filter and dual-media filter in filtration mode and in backwashing mode. For a color version of this figure, see www.iste.co.uk/gaid/reuse1.zip*

Table 7.1 shows the maximum acceptable turbidity for different types of filters and the average cycle time. The type of gravity filter is selected according to the particle load and the treatment objectives.

Filter type	Maximum acceptable SS ($mg{\cdot}L^{-1}$)	Average cycle time (h)
Rapid sand filter	5–10	24
High rate sand filter	5–30	24–48
High rate dual-media filter	5–30	24–48

Table 7.1. *Maximum acceptable turbidity for different types of filters*

7.1.3. *Pressure filters*

Pressure filtration is carried out in closed cylindrical vessels, working generally at an absolute pressure of 3–5 bar. The filtration rates implemented are 8–15 $m{\cdot}h^{-1}$.

Whether single media or dual media, the same materials used in gravity filtration are included in pressure filters.

Pressure filters most often require an intermediate water pumping at the end of the secondary clarifier. Particle retention mechanisms and pressure drop development, as well as pretreatments, are analogous to those of open filter gravity filtration.

The filter is backwashed, after atmospheric pressurization. The backwash sequences are similar to those of a gravity flow sand filter.

The main advantage of a pressurized filter is that it has a very high maximum available load, which allows for longer cycle times.

Pressure filters are backwashed in three phases: air, air/water and rinse. They are similar to those of a gravity flow sand filter.

During backwashing, the volume expansion is between 20 and 30% by means of a wash flow for a single-media sand filter of 25 $m^3{\cdot}m^{-2}{\cdot}h^{-1}$. The air/water wash times must be well controlled, as the water retention above the media is lower than in gravity filters.

The backwash process often results in sand loss, although it is not noticeable significantly, thus requiring periodic sand replenishment.

7.2. Gravity filtration with reagents

Tertiary filtration, combined with an aluminum or iron coagulant, has the advantage of ensuring better turbidity at the outlet and obtaining a better water quality from a microbiological point of view. This is due to the elimination of small particles that are trapped by the flocs formed by the chemical reagent.

As for drinking water, the good retention of particles is based on the formation of a hydroxide floc that traps impurities, promotes the agglomeration of particles by facilitating the formation of a chemical or ionic bond with thc filter media, accumulates within the filter and allows to deliver water of better physical and microbiological quality. Particle retention is based on physical and chemical parameters.

In the absence of coagulant, microbial reduction is less high. The retention of small particles is now dependent only on the media characteristics and operating conditions (filtration rate, media height and cycle time). Particle retention is only based on physical parameters.

The improved microbiological quality of wastewater comes with reduced BOD, SS and phosphorus.

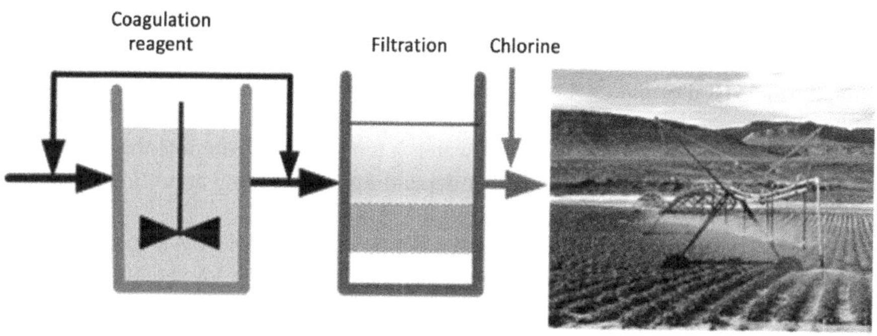

Figure 7.5. *Direct filtration with coagulation. For a color version of this figure, see www.iste.co.uk/gaid/reuse1.zip*

In direct filtration with reagent, instrumentation is very important for automatic control of the treatment. Monitoring of the water resource is also imperative to alert operators to variations and allow them to adjust reagents.

At low temperatures, coagulants such as polyaluminum chloride are preferred over aluminum and iron salts as $AlCl_3$ are already hydrolyzed, and the temperature has less effect on the coagulation/flocculation process within the filter.

This solution ensures better elimination of microorganisms, which improves the effectiveness of disinfection. However, it has the disadvantage of discharging sludge (after backwashing the filters), which can be recirculated upstream of the wastewater treatment plant (WWTP). It involves an additional cost for reagents. The dosage of reagents (Al or Fe) ranges from 1.0 to 3.0 $g \cdot m^{-3}$.

Gravity filtration with reagents can also be dimensioned for a dual-media filter with an Al- or Fe-based coagulant. Dosages can vary between 0.5 and 4 g metal$\cdot m^{-3}$ depending on the quality of the wastewater to be treated.

7.3. Filtration mechanisms

Tertiary wastewater filtration appears to be an interesting choice among other alternatives as a solution for refining treated wastewater due to its simple implementation and exploitation while producing a water free of or low in SS.

However, filter retention mechanisms are sensitive to variations in SS of clarified water. The higher their concentration, the shorter the cycle times and the more variable the quality of the filtered water. In the presence of residual SS in filtered water, BOD5 and COD will be higher as they include organic particles.

Due to the organic content of wastewater, a tertiary filter can potentially function as a physicochemical or biological filter depending on operating conditions.

Filtration can be defined as the interception of a particle by a material grain called a collector. The transport of suspended particles in the immediate vicinity of the filter grains depends on physical and hydraulic factors including flow rate, size and shape of the filter media, filter configuration, filter porosity and finally the physical characteristics of the suspended particles (shape, size and density).

If we consider particles as spheres, the hydrodynamic force acting on them includes a hydrodynamic extraction force and a hydrodynamic driving force, which has the role of rolling the particles along the material grains.

The effect of the hydrodynamic force is therefore more important in wastewater filtration, where there is usually a much larger proportion of large particles to be filtered, whereas this is not the case in drinking water filtration.

The hydrodynamic driving force increases with filtration rate. Particles are separated from the collector when the tangential force is greater than the fixing forces between the particle and the collector.

At low rate filtration (<10 $m \cdot h^{-1}$), the tangential force is positive for all particle sizes so that scouring is low.

The filtration process is a two-step process. The first is called the "initial step", which is the introduction of particles into a clean filter. Retention occurs on clean material grains. The second step is a transitional step during which thc particle deposition occurs on the surface of the material grains, which are then covered by previously deposited particles. During this period, particle retention improves or degrades if the chemical characteristics of the particles have been modified in terms of electrical charges.

At the end of the cycle, the saturation of the filtering material and the hydrodynamic conditions tend to degrade the retention and, on the contrary, induce filter breakthrough by particle leakage. Particles and material grains have surface charges that may be different.

If they have opposite charges, filter retention conditions are favorable due to their mutual attraction. If their charges have the same sign, filter retention conditions are unfavorable due to repulsive interactions.

The filter performance is effective if the aggregation and deposition of suspended particles are optimal and promoted by particle/particle and media/particle interactions, which are optimal at pH 6.5–7.5.

Changing the pH can alter the wastewater matrix by reacting with H^+ and OH^- and affect the adhesion of particles to materials grains. One of the most important control parameters for solids retention in depth filters is the electric double layer (EDL) and the London–Van der Waals forces.

However, due to the various transport mechanisms of the particles, they move through the circulation currents and end up next to the surface of a grain. At this point, adhesion forces must be present for the particles to be retained on the material grains or on the particles that have already been deposited. If the deposited particles are carried away by the circulating water, a detachment mechanism is at work. Particle retention therefore depends on surface forces (Van der Waals forces) between particles and material grains when the distance separating them is of the order of the nanometer.

Particles are usually smaller than interstices. As a result, they follow the water circulation currents; many of them should not touch the surface of the material grain and should end up in the filtered water.

But due to the various particle transport mechanisms, they move through the circulation currents and arrive next to the surface of a grain. At that moment, there must be some fixing forces to hold the particles on the material grains or on the already deposited particles. If the deposited particles are driven by circulating water, this means that a detachment mechanism is in place.

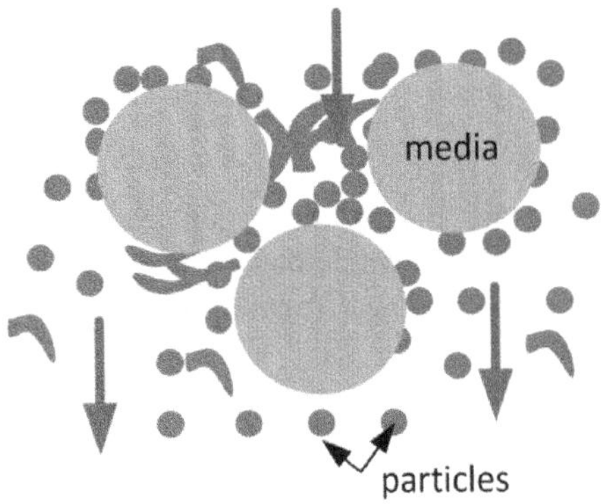

Figure 7.6. *Particle attachment and detachment mechanisms. For a color version of this figure, see www.iste.co.uk/gaid/reuse1.zip*

7.4. Implementation parameters

7.4.1. *Media*

The mechanical filtration media used must be insoluble, non-friable and free of any substances that could impair water qualities. The filter media are characterized by their grain size distribution, particle shape, specific mass and porosity, while suspension particles are mainly described by their size and concentration.

Media depend on the type of filter used (single media and dual media) and the purpose of the treatment. The various media used in wastewater include quartz sand, gravel, anthracite, pumice, expanded shale or pozzolan (Biodagene, volcanic stones). Other materials such as Filtralite (expanded clay) are available on the market.

For each filter media, the physical characteristics determined in the laboratory or indicated by the supplier are:

– ES d_{10} (opening allowing passage of 10% (w/w) – particles) in mm;

– the uniformity coefficient (UC) (d_{60}/d_{10});

– the maximum grain size d_{90} in mm;

– the solid density (absolute), dry density (apparent in the air), wet density (saturated with water) and porosity (g·cm^{-3});

– the hardness (Mhos);

– the geometric shape (spherical, ellipsoidal, granular, cylindrical, irregular, angular);

– the composition, determined, as appropriate, by the degree of purity (silica content, etc.), solubility in acid or water, ash content (activated charcoal), etc.

The equivalent unit in English-speaking countries is mesh (US or BS with the following conversion: size in mesh = 15/size in mm).

Table 7.2 summarizes the characteristics of various materials used in wastewater treatment.

Unlike drinking water, wastewater filtration on granular media is a complex process as the efficiency of the process depends on many interdependent variables.

Materials	Physical characteristics						
	ES d_{10} (mm)	UC (d_{60}/d_{10})	d_{90} (mm)	Hardness (Mhos)	Porosity, ε	Absolute density (g·cm^{-3})	Bulk density (g·cm^{-3})
Gravel	5	<2	9.00	–	–	2.65	1.65
Fine sand	0.85	≤1.3	1.25	–	0.45	2.65	1.50
Average sand	1.0	≤1.3	1.55	–	0.45	2.65	1.50
Coarse sand	1.35	≤1.45	2.10	–	0.45	2.65	1.50
Anthracite	1.4	<1.50	2.50	>4	0.55	1.45–1.60	0.85–0.95
Pumice	1.2–1.3	<1.45	2.00	>4	0.50–0.52	0.8–1.2	1.30–1.55
Schist (Biodagene)	2.5	<1.7	5.00	>4.5	0.40	1.60	0.96

Table 7.2. *Characteristics of the various filtration materials*

The most important design factors are the characteristics of the filter media, including the type of filter media, grain size, properties of the materials to be filtered and the applied load.

Single-media filtration is widely used in tertiary treatment of wastewater, while dual-media filtration is gradually expanding. Single-media filtration is often preferred because of its simplicity of construction. High-height rapid filters typically involve sand heights up to 2.0 m with an ES ranging from 1.25 to 2.5 mm and a UC of 1.2–1.5.

The rates passing through gravity single-media filters are variable between 5 and 20 $m \cdot h^{-1}$.

The size of tertiary filters depends on the material, water quality to be filtered, upstream processes, quality objectives for filtered water, cleaning method, filtration rate and duration of filtration cycles. The sand used in filtration is a natural material, based on silica, from rivers, natural deposits, dunes, etc. It is proposed in two forms:

– *rolled sand* (alluvial or loose rock), whose rounded shape was acquired by erosion, is a natural sand, sifted after washing and drying. It is used in or near water courses or in shallow seabed;

– *crushed sand* is obtained by felling and crushing massive rocks, which give them angular shapes favorable to the particle retention during filtration. This sand has a low sphericity, but better porosity.

Anthracite is a very hard mineral coal characterized by its high carbon content, which can reach up to 95%. It comes in the form of hard and angular grains. It is laid in a top layer with a grain size large enough to retain particles of different sizes.

Light materials such as *pumice* (light grey volcanic lava made very light by numerous gas bubbles trapped during cooling), pozzolan (volcanic rock with a brown-black to brick red color and a honeycomb structure), expanded clay or shale are used in filtration and as a support for biological development.

The support *gravel* is not a filtration material. It retains the filter material, which is placed above it, improves water distribution during washing and protects nozzles during washing. Its characteristics depend on those of the filter floor and material.

For all these materials, an ES is defined, d_{10}, which represents the size of the opening through which 10% by weight of material passes.

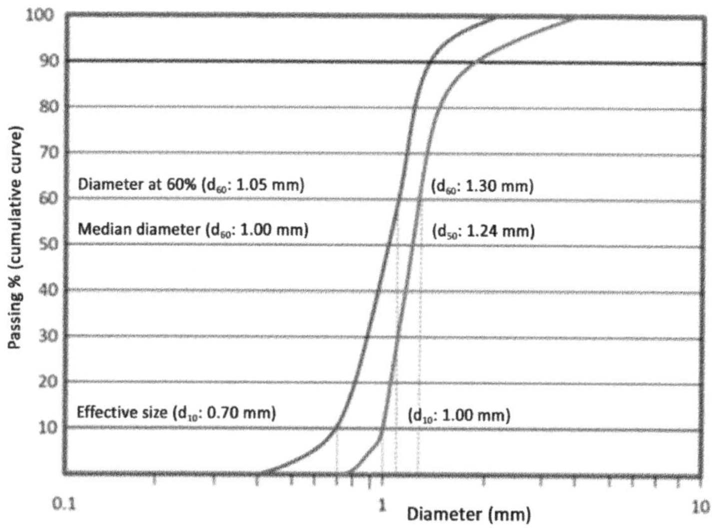

Figure 7.7. *Cumulative distribution curve of sand grain diameters (d_{10}: 0.70 mm and d_{10}: 1.00 mm). For a color version of this figure, see www.iste.co.uk/gaid/reuse1.zip*

The diameter d_{50} is calculated by the formula: $d_{50} = \text{ES. UC}^{0.834}$.

The diameter d_{90} is given by: $d_{90} = d_{10}\ 10^{1.67 \log \text{UC}}$.

Commercial filter media are usually specified by their ES and the UC, calculated according to the ratio d_{60}/d_{10}.

The smaller the ES, the larger the particle adhesion surface, but the higher the pressure loss and retention is in the first half of the filter.

The UC is the ratio of the diameter that allows 60% of particles to pass and the diameter that allows 10% of particles to pass %.

Namely:

$$\text{UC} = \frac{d_{60}}{d_{10}}$$

d_{60} is the size of the opening through which 60% of the sand passes. This coefficient gives an indication of the grain size homogeneity of the filter mass. The UC values must be less than 1.6 and are generally between 1.3 and 1.4. Lower UC values make the material expensive, as a high proportion of fine material is lost during washing and higher values would reduce material porosity.

Porosity ε is defined as the fraction of empty volume relative to the total volume of material. It represents the void ratio between the stacked material grains, so the volume occupied by the material is $(1 - \varepsilon)$. Depending on the nature of the porous medium, ε varies between values close from 0 to 1.

The void is defined by:

$$\varepsilon/(1 - \varepsilon)$$

Thus, the higher ε is, the lower the resistance to water flow. Retention is partly in voids that slow down the flow of water as solids settle into the filter.

The porosity of a material does not depend on its size, but only on its shape and geometric characteristics. The porosity of sand, for example, is between 42% and 45%, while that of anthracite is 50–55%.

The weight loss to fire at 450°C must be less than 2% and the weight loss to hydrochloric acid washing (20% v/v for 24 h at 20°C) must be less than 2%.

The apparent density is the ratio of the material mass divided by the total volume occupied. It is therefore the mass of dry bulk sand occupying the volume unit, ranging from 1,400 to 1,600 $kg \cdot m^{-3}$ for rolled silica-calcareous sands.

In reality, the apparent density is a dimensionless parameter, as it is the ratio of the density to mass of the same volume of water (at a given temperature). It is equal to one thousandth of the density. The absolute density of the material, that is to say the volume of voids excluded, is much higher. It is equal to 2,500–2,600 $kg \cdot m^{-3}$ for the same rolled silica-calcareous sands. When they are in the same filter, the specific mass of the materials is useful to reclassify them after backwashing the filter.

Grain size determines the specific surface of a material, which is the sum of all outer surfaces of the particles contained in a unit mass.

Thus, for a material grain of mass M and density ρ, the equivalent volume V is given by:

$$V = M/\rho$$

The equivalent diameter (d) of the sphere can be calculated with:

$$V = \pi d^3/6$$

The specific surface of a material as (m^{-1}) is the surface developed by a grain reduced to the volume of that grain.

For a spherical particle, the specific area is:

$$\frac{Ap}{Vp} = \left(\frac{\pi d^2}{\frac{\pi d^3}{6}}\right) = \frac{6}{d}$$

where Ap is the grain surface (m^2), Vp is the volume (m^3) and d is the diameter (m).

For non-spherical grains, the form factor ψ is introduced.

Namely:

$$\frac{Ap}{Vp} = \frac{6}{\psi d}$$

The sphericity factor of a grain measures the similarity of its form with the form of a sphere:

$$\psi = \frac{\text{Grain volume}}{\text{Circumscribed sphere volume}}$$

If the grain is almost spherical, it fills almost the entire circumscribed sphere. If, instead, the grain is more like a thin disk, it occupies a small piece inside the sphere. The sphericity increases as the grain takes up more and more space. The sphericity factor is 1 when both volumes coincide. If a grain is not spherical, it has more surface area for the same volume and therefore offers more resistance to transport.

The grain size distribution and the sphericity (ψ) dictate the arrangement of materials in two-layer filters. Porosity affects hydrodynamics during filtration cycles and the washing process.

A material with an inadequate sphericity may encourage preferential flow paths through the filter. The sphericity is therefore an important parameter when choosing filter materials.

The sphericity coefficient ψ represents the ratio of a sphere surface area to the particle surface area. A perfect sphere has a sphericity coefficient equal to 1 while the materials sold for filtration have values between 0.5 and 0.8.

Several formulas are proposed to determine the coefficient ψ:

$$\psi = \frac{\sqrt{d_{10}}}{\sqrt[3]{d_{90}d_{60}}}$$

$$\psi = \frac{d_{10}}{d_{90}}$$

$$\psi = \frac{2(\sqrt{\pi A})}{P}$$

P is the grain perimeter (m).

The hardness of the material is defined by its resistance to attrition for its filtration application. The hardness measurement on the Mohs scale depends on the ability of a material to scratch a reference material or to be scratched by a reference material. The measurement is based on a scale from 0 to 10, as the latter value corresponds to diamond, which can only be scratched by another diamond. The hardness should preferably be greater than 3 to ensure that the washing operations do not lead to a crumbling of the material.

The filter coefficient and the Kozeny pressure drop expression are calculated using the following formulae:

$$\lambda = \frac{3(1-\varepsilon)\eta\alpha}{2d}$$

$$\frac{\Delta p}{H} = \frac{K\,\mu}{\rho_{water}\, g\, \varepsilon^3}\left(\frac{6(1-\varepsilon)}{d}\right)^2 v$$

where λ is the filter coefficient in m^{-1}, d is the diameter of the media in m, Δp is the pressure drop in Pa, H is the media height in m, μ is the dynamic viscosity in Ns m^{-2}, ρ_{water} is the water density mass in kg m^{-3}, g is the acceleration due to the gravity in m s^{-2}, v is the filtration rate in $m{\cdot}s^{-1}$, ε is the filter porosity, α is the retention efficiency, η is the transport coefficient, K is the Kozeny coefficient and d is the diameter of the spherical medium in m.

In addition to the characteristics of the filter media, the structure of the filter determines its performance. The media dimensions are generally larger than those of drinking water filters to minimize pressure drop through the filter, as the clarified wastewater has concentrations in SS between 10 and 35 $mg{\cdot}L^{-1}$ while drinking water subject to coagulation–flocculation–settling has an SS concentration of less than 5 $mg{\cdot}L^{-1}$.

7.4.2. *Media height ratio/d_{10}*

The objective of this ratio is to maintain a constant filtered water quality for a particle input concentration and a filtration rate not varying much. This H/d_{10} ratio maintains consistent particle elimination performance as the filtration height and material grain diameter increase.

For drinking water, one of the filter size parameters recommends that the H/d_{10} ratio (media height H and size d_{10}) be between 1,000 and 1,200. However, as wastewater has a higher concentration of SS, the quality of treated water is generally lower than that required for drinking water; tertiary wastewater filters use a coarser medium to facilitate operation. For drinking water, it is often required that the output turbidity be equal to or less than 0.5 NTU while for wastewater, the turbidity must be less than 5 NTU. As such, the H/d_{10} ratio is recommended with a variable range between 750 and 1,500 due to fluctuations in the filtration process and broader quality guarantees that are required.

A ratio between the height of the filter layer (H) and the grain diameter (d_{10}) must be respected. For a sand filter, the H/d_{10} ratio must be greater than 1,000. For two-layer filters, this ratio must be greater than 1,200. In the case of multimedia filters, the L/d ratio is calculated for each media and then added to give the total filter ratio. Gravel is not counted. The following ratios are thus obtained for dual-media filters:

$$\frac{(H_1)}{(d_{10})1} + \frac{(H_2)}{(d_{10})2} > 1{,}200$$

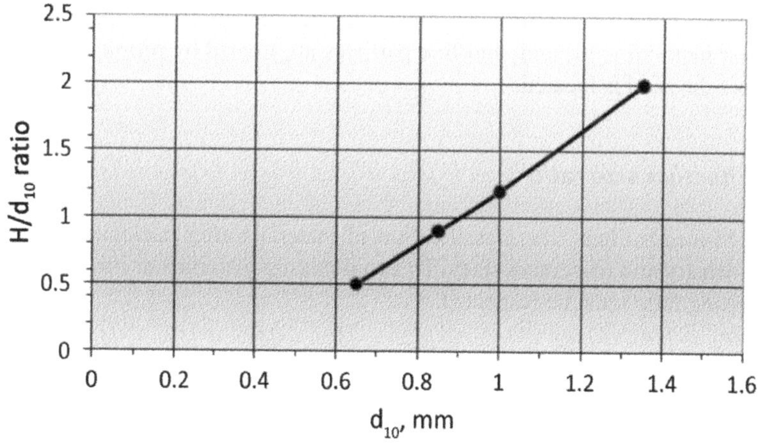

Figure 7.8. *Ratio H/d_{10} for sand filters*

For deep filtration applicable to HRs filters, with an ES greater than 1.35 mm, the H/d_{10} ratio should be greater than 1,400.

Many filters are designed with total media heights between 1 and 2 m. In dual-media filters, the d_1/d_2 and H_1/H_2 ratios are essential to achieve good particle retention, output turbidity refining and easy reclassification after backwashing.

Table 7.3 shows some typical cases used in treatment processes.

Filter type	Materials	Total height (mm)	Height (mm)	d_{10} (mm)	UC	H/d_{10}	Recommended value
Rapid single-media filter	Sand	1,000	1,000	0.85	1.3	1,176	>1,000
HR single-media filter	Sand	2,000	2,000	1.35	<1.4	1,481	>1,400
Rapid dual-media filter	Anthracite or pumice Sand	1,400–1,600	700–800 700–800	1.20 0.85	<1.4 1.3	1,323–1,512 –	>1,200 –
HR dual-media filter	Anthracite or pumice Sand	2,000	1,500 500	1.40 0.65	<1.4 1.3	1,840 –	>1,400

Table 7.3. *Examples of H/d₁₀ ratios for different types of filters*

During wastewater filtration, large particles accumulate in the upper part where the largest material is located, and fine particles are treated by refining in the lower part where the sand is located.

7.4.3. *Effective size ratio*

In dual-media filters, the classification of materials after backwashing depends on their density and respective sizes. To avoid mixing materials or misclassification, the following ratio must be respected:

$$\frac{d1}{d2} = \left(\frac{\rho 2-\rho_{water}}{\rho 1-\rho_{water}}\right)^{0.667}$$

d_1 and d_2 are the d_{10} of each media, and ρ_1 and ρ_2 are the density of both media. ρ_{water} is the water density. For example, in an anthracite/sand dual-media filter, the ES of the anthracite can be deduced from the sand's ES.

If $d_2 = d_{10}$ of sand: 0.85 mm, $\rho_2 = \rho_{sand} = 2.65$ g·cm^{-3} and $\rho_1 = \rho_{anthracite} = 1.55$ g·cm^{-3}. This is deducted for $\rho_{water} = 1$ g·cm^{-3}:

$$d1 = d2\left(\frac{\rho 2 - \rho_{water}}{\rho 1 - \rho_{water}}\right)^{0.667} = 0.85\left(\frac{2.65 - 1}{1.55 - 1}\right)^{0.667} = 1.76 \text{ mm}$$

To select the ESs of media in dual-media filters, the ratio between the ESs and the density of the media and water is used:

$$\frac{d1}{d2} = \left(\frac{(\rho 2 - \rho_{water})}{(\rho 1 - \rho_{water})}\right)^{0.667}$$

Index 1 refers to the finest material and index 2 refers to the largest material. For example, in a sand/anthracite dual-media filter, the ratio d_1/d_2 corresponds to the ratio between the ES of the sand and that of the anthracite. This equation is used to avoid mixing layers during backwashing. Indeed, if the ratio between the ES of both materials is insufficient, the media of the lower layer could be mixed with the media of the upper layer when the filter bed is expanding.

For anthracite/sand dual-media filters, a ratio between 1.75 and 3.0 is recommended.

7.5. Size parameters: filtration rate and media height

Filtration is made in a downstream direction at a rate equal to:

$$V = Q/S$$

where V is expressed in m·h^{-1}, Q is expressed in m^3·h^{-1} and S is expressed in m^2.

The application of a high filtration rate implies a higher media height, a larger effective grain size and a significant increase in the backwashing rate.

The use of a larger grain size and higher filtration rate results in an increase in SS penetration depth without any decrease in cycle time or quality of filtered water.

Filtration rates vary between 6 and 8 m·h^{-1} for a sand filter (0.9–1.2 m), 14 and 16 m·h^{-1} for an HR filter made of expanded shale, 12 and 15 m·h^{-1} for a rapid dual-media filter, and 15 and 20 m·h^{-1} for a dual-media filter operating at HR.

High filtration rates are applicable in single-media or dual-media filters, depending on the characteristics of the filter media and the total height of the media.

Filter type	Filtration rate $m \cdot h^{-1}$	Media total height m
Rapid sand filter	6–8	1.0–1.2
HR sand filter	14–16	2
Rapid dual-media filter	12–15	1.4–1.6
HR dual-media filter	15–20	2.0

Table 7.4. *Filtration rates and media heights for different filters*

The high filtration rate increases particle diffusion within the filter mass and the hydrodynamic forces. The dual-media filter has a structure including the smallest media laid in the lower layer of the filter, thus preserving the capacity to trap fine particles.

This gives the dual-media filter a capacity to handle a large flow rate and ensure higher SS retention, as the larger grain size of the material laid in the upper layer of the filter allows for greater storage of solids.

In addition, the accumulation of solids in the top layer tends to fill the gaps (interstices) between the grains and refine the quality of the filtered water.

7.5.1. *Filter depth modelling*

One of the relationships that govern the tertiary filtration mechanism is:

$$Cs = Co \exp(-\lambda H)$$

$$\lambda = \frac{3(1-\varepsilon)\eta\alpha}{2d10}$$

where Co and Cs are the SS concentrations at the inlet and outlet of the filters ($mg \cdot L^{-1}$), H is the height of media (m), λ is the filter coefficient in unit m^{-1}, d_{10} is the size of the media (m), ε is porosity, α is the fixing efficiency and η is the transport coefficient. By replacing λ in the expression of Cs, we can deduce H, the filtering height:

$$1 - \frac{Cs}{Co} = 1 - \exp(-\frac{3}{2}(1-\varepsilon)\,\alpha\eta(\frac{H}{d_{10}})) = 0.80$$

For example, for turbidity expected at the outlet equal to 5 NTU (SS < 10 mg·L^{-1}) for a filter outlet value equal to 25 NTU or approximately 37 mg·L^{-1} of SS:

$$\frac{3}{2}(1-\varepsilon)\,\alpha\eta\left(\frac{H}{d_{10}}\right) = Ln(0.20) = 1.60$$

For $\alpha = 1$; $\eta = 0.0015$ mm, $\varepsilon = 0.51$, we deduce $H/d_{10} = 1{,}450$.

Deeper filters enable better retention of solids and better turbidity at the outlet, since the accumulation of solids within the filter mass tends to trap fine particles which are titrated in turbidity. The aim is to find a compromise between cycle time (≥24 h), media height (around 2 m), filtration rate (around 20 m·h^{-1}), media grain size (d_{10}: 1.35 mm), pressure loss and turbidity at the outlet (≥5 NTU). A deep filter improves the quality of the effluent, combined with a large retention of solids.

At conventional filtration rates for wastewater (<10 m·h^{-1}), the initial layers of the filter retain solids as the rate is not strong enough to diffuse solids deep. The retention of solids decreases with depth.

For this purpose, tertiary filters operating at these rates have a sand height of 0.9–1.1 m with a d_{10} grain size of 0.85–0.90 mm. HR filtration is a process that affects both the retention and detachment of particles. Particle retention is calculated from the transport efficiency η and the attachment efficiency α. A HR rate can increase the hydrodynamic extraction force without undermining the overall retention of particles in the filter. In two-layer filters, the bottom material is small enough to retain the solid front moving along the filter and produce low turbidity water. Hydrodynamic forces are therefore different in the dual media of the filter.

The use of coarse media has, however, a disadvantage in that it reduces the particle capture area. Dual-media filters benefit from the use of large grains with low or medium density and smaller grains with higher density. Such a design allows the backwashing to stratify the filter bed, by keeping large grains at the top and small grains at the bottom, thus encouraging penetration in depth of the solids while maintaining the desired performance in terms of treated water quality.

Note that 80% of the filter height is used for storage of solids when the filter is running at HR. For conventional rates, less than 50% of the height is used. Using the entire filter depth for solid storage ensures slow development of pressure drop.

7.6. Operating parameters

7.6.1. *Pressure drop in a clean filter*

The allowable pressure drop is equal to the difference between the liquid level in the filter during operation and the restitution (retention) level in the basin.

It includes:

– the pressure drop through the clean filter;

– the pressure loss due to clogging and control equipment;

– the pressure loss due to the filter floor nozzles.

The calculation of pressure drop in a clean filter includes the consideration of several parameters such as filtration rate, filtration height and filter material characteristics. The equation developed by Carman–Kozeny for a single-media filter is:

$$\Delta P = \frac{Kv\,(1-\varepsilon)^2}{\rho\, g\, d^2p\, \varepsilon^3}\,\mu\, H\, v + \frac{Ki\,(1-\varepsilon)\, H\, v^2}{g\, dp\, \varepsilon^3}$$

with:

– ΔP: pressure drop (m water);

– H: filtration height (m);

– ε: porosity of clean filter;

– dp: d_{10} of filter media (m);

– μ: water dynamic viscosity ($kg \cdot m^{-1} \cdot s^{-1}$);

– v: filtration rate ($m \cdot s^{-1}$);

– ρ: water density ($kg \cdot m^{-3}$) for a given temperature (°C);

– g: gravity acceleration ($m \cdot s^{-2}$);

– Kv: coefficient due to viscous forces;

– Ki: coefficient due to inertial forces.

Table 7.5 indicates the values of Kv, Ki and ε for sand and anthracite.

Filter media	Kv	Ki	ε
Sand	150	1.75	0.45
Anthracite	228	4.4	0.55

Table 7.5. *Pressure drop calculation parameters*

Thus, for a sand filter 1 m high, operating at a speed of 7 m·h^{-1} (0.00192 m·s^{-1}), the pressure drop in clean water is equal to:

$$\Delta P = \frac{150\,(1-0.45)^2}{1000.9.81(0.001^2)(0.45^3)}\,0.00114.1.0.00192 + \frac{1.75.(1-0.45).1.(0.00192^2)}{9.81.0.001.(0.45^3)}$$

ΔP = 0.12 m

with: ε = 0.45; dp = 0.001 m; μ = 0.00114 kg·m^{-1}·s^{-1}; ρ = 1,000 kg·m^{-3}; g = 9.81 m·s^{-2}; Kv = 150 and Ki = 1.75.

Figure 7.9 shows the evolution of Δ p/H depending on the filtration rate and for different effective sand sizes. For a dual-media filter, the sum of the respective pressure losses of both media should be calculated. Thus, we have:

$$\Delta P1\ (anthracite) = \frac{Kv1\,(1-\varepsilon 1)^2}{\rho\, g\, d^2 p1\, \varepsilon 1^3}\,\mu\, H\, v + \frac{Ki1\,(1-\varepsilon 1)\,H\,v^2}{g\, dp1\, \varepsilon 1^3}$$

$$\Delta P2\ (sand) = \frac{Kv2\,(1-\varepsilon 2)^2}{\rho\, g\, d^2 p2\, \varepsilon 2^3}\,\mu\, H\, v + \frac{Ki2\,(1-\varepsilon 2)\,H\,v^2}{g\, dp2\, \varepsilon 2^3}$$

The total clean filter pressure drop is the sum of $\Delta p_1 + \Delta p_2$, using the respective coefficients Kv and Ki for each media.

Thus, for a filter of 1.6 m in total height composed of 0.8 m of anthracite and 0.8 m of sand, operating at a rate of 12 m·h^{-1} (0.00333 m·s^{-1}), the pressure drop in a clean filter is equal to:

$$\Delta P1 = \frac{228\,(1-0.55)^2}{1000.9.81(0.0014^2)(0.55^3)}\,0.00114.0.8.0.00333 + \frac{4.4\,(1-0.55).0.8.(0.00333^2)}{9.81.0.0014.(0.55^3)}$$

$$\Delta P2 = \frac{150\,(1-0.45)^2}{1000.9.81(0.00085^2)(0.45^3)}\,0.00114.0.8.0.00333 + \frac{1.75(1-0.45).0.8.(0.00333^2)}{9.81.0.00085.(0.45^3)}$$

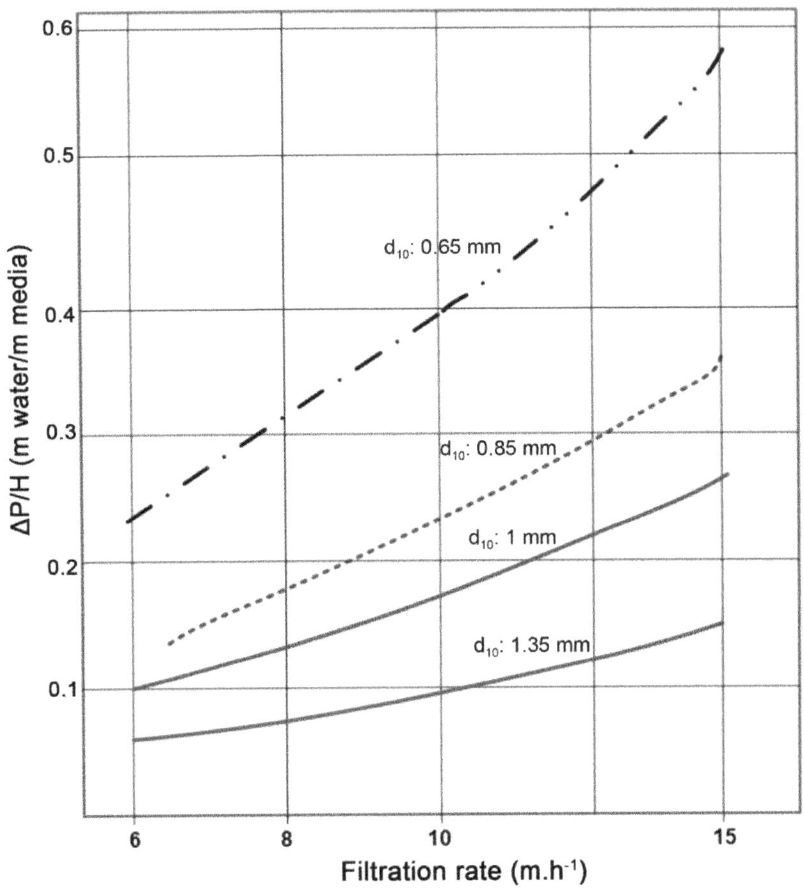

Figure 7.9. *ΔP/H obtained for different effective sand sizes and different filtration rates*

Namely, $\Delta P = \Delta P1 + \Delta P2 = 0.05 + 0.22 = 0.27$ m.

With: $\varepsilon = 0.55$; dp $= 0.0014$ m; $\mu = 0.00114$ kg·m^{-1}s^{-1}; $\rho = 1{,}000$ kg·m^{-3}; g = 9.81 m·s^{-2}; Kv = 228 and Ki = 4.4 for anthracite.

And: $\varepsilon = 0.45$; dp $= 0.00085$ m; $\mu = 0.00114$ kg·m^{-1}s^{-1}; $\rho = 1{,}000$ kg·m^{-1}s^{-1}; g = 9.81 m·s^{-2}; Kv = 150 and Ki = 1.75 for sand.

7.6.2. *Retention capacity*

It represents the quantity of material that the filter is likely to retain during a filtration cycle. It is usually expressed in kg of media per m^2 of media for rapid filters and in $kg \cdot m^{-3}$ for HRFs.

The retention of solids in the filter depends on both the interactions of particles with the filter media and the solids that settle within the filter mass. Unlike drinking water, the wastewater matrix is different as it contains solid particles with a high biological content that behaves like a sticky material on materials. In addition, there may be large diurnal and seasonal variations in performance that can affect the operation of filters. The SS retention takes place mainly in the upper part of the filter, even if it extends to the second medium, therefore using the entire depth of the filter to retain the particles.

The retention capacity of a filter depends on the porosity of the material. In a clogging situation, it can be assumed that the interstices form an internal area itself inscribed in a circle. Each arc of this area is calculated as follows:

$$\text{arc A} = 0.0174 \times 60 \times (d_{10}/2) \text{ with an angle of } 60°$$

This area $A_1A_2A_3$ (Figure 7.10) has an area equal to:

$$S = 0.25 \sqrt{[(3 \times \text{arc}) \times (\text{arc } A_1) \times (\text{arc } A_2) \times (\text{arc } A_3)]}$$

knowing that the arcs A_1, A_2 and A_3 are equal.

This area perimeter is given by:

$$P = 0.5 \times (3 \times \text{arc A})$$

Th radius of the circle inscribed in this area is Rz = S/P.

The inscribed circle area is: π Rz 2. The diameter is therefore equal to 2Rz. This shows (see example below) that the inscribed circle diameter is equal to 30% of d_{10} of the material.

The volume of pores available in the filter is given by: Vp = S.H.ε for an area S and a height H. The volume available is therefore 30% of the volume of the pores, that is Vd = 0.30 Vp.

For a particle concentration in the interstices varying between 15 and 20 $g \cdot L^{-1}$ according to the type of floc (hydroxide floc, silt, coarse particles, etc.), the theoretical retention capacity is deduced from this ($kg \cdot m^{-2}$).

Parameters	Calculations $d_{10} = 0.85$ mm	Calculations $d_{10} = 1.35$ mm
Arc A	0.0174 × 60 × (0.85/2) = 0.443	0.704
S (mm²)	0.25 √[(3 × 0.443) × (0.443) × (0.443) × (0.443)] = 0.0852	0.215
Perimeter (mm)	0.5 × (3 × 0.443) = 0.665	1.057
Inscribed circle radius (mm)	0.0852/0.665 = 0.128	0.203
Diameter (mm)	2 × 0.128 = 0.256	0.406
Inscribed circle area (mm²)	3.14 × 0.128² = 0.0515	0.129
Inscribed diameter ratio/d_{10}	0.256/0.85 = 0.30	0.406/1.35 = 0.30

Table 7.6. *Calculation examples of the interstice diameter ratio/d_{10} of the material*

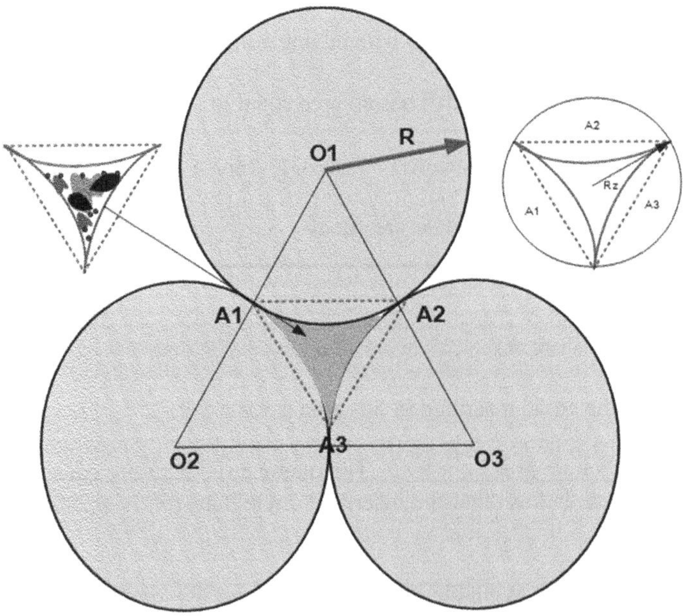

Figure 7.10. *Interstices of sand grains saturated in retained particles. For a color version of this figure, see www.iste.co.uk/gaid/reuse1.zip*

The actual capacity of gravity sand filters operating at speeds ≤ 8 m·h^{-1} for a height <1.20 m is about 40% of the theoretical capacity, as the clogging is established in the first 40 cm of material height. This accumulation results in a rapid increase of pressure loss, although there may still be a slow diffusion of the particle front toward the lower areas of the filter.

Sand		**Shale**	
Parameter	**Calculations**	**Parameter**	**Calculations**
Media height (m)	1	Media height (m)	2
Area (m^2)	1	Area (m^2)	1
E	0.45	E	0.40
Pore volume (interstices) mm^3	1 × 1 × 0.45 = 0.45	Pore volume (interstices) mm^3	2 × 1 × 0.45 = 0.90
Available volume (m^3) (0.30: ratio $d_{interstice}/d_{10}$)	0.45 × 0.30 = 0.135	Available volume (m^3) (0.30: ratio $d_{interstice}/d_{10}$)	0.90 × 0.30 = 0.27
Floc concentration in interstices (kg·m^{-3})	15–20	Floc concentration in interstices (kg·m^{-3})	20–30
Theoretical floc capacity over the entire height (kg·m^{-2})	0.135 × 15 = 2.025	Theoretical retention capacity over the entire height (kg·m^{-2})	0.27 × 20 = 5.40
Actual retention capacity, that is, 40% of the height (kg·m^{-2})	2.025 × 0.4 = 0.81	Actual retention capacity, that is, 70% of the height (kg·m^{-3})	5.40 × 0.70 = 3.78

Table 7.7. *Calculation of retention capacity of sand filters*

In a HR filter containing a large material such as schist, clogging is done in depth as the HR tends to push flocs toward the lower parts of the filter, which explains that the capacity is about 70% and is expressed in kg·m^{-3}.

For dual-media filters (anthracite/sand or pumice/sand), an identical calculation can be made by adding up the capacities calculated for both media. However, as the porosity of anthracite is 0.55, the total pore volume is used as it is estimated that the particle diffusion occurs over the entire height of this media.

7.6.3. *Filter backwashing conditions*

7.6.3.1. *Theory and calculations*

The basic purpose of backwashing is to detach the maximum number of particles accumulated during a filtration cycle and to remove them from the filter. The filter then starts a new cycle under favorable pressure drop conditions. The main challenge of washing is to achieve the highest possible particle detachment rate. Proper backwashing is a very important step in the operation of a filter. If the filter is not properly washed, operational problems arise, such as too short cycles, high water losses, lower output, too high outlet turbidity, etc.

The filter must be washed when the following conditions are met:

– the pressure drop is so high that the filter no longer produces water at the desired flow rate;

– or flocs begin to pierce the filter and turbidity in the filtered water increases;

– or the filtration cycle reaches its expected operating time.

If the water feeding a filter is of very good quality, filtration cycles can be very long. Some filters may work longer than a week before needing to be washed. However, this is not recommended as long filtration cycles can cause the filter material to compact, so it is difficult to expand the media height during washing.

Particle detachment requires a higher energy thrust than the forces that bind them to the material. The detachment rate depends on both the attachment forces and the forces enabling detachment. All these forces depend on the particle structure and composition. Especially since the profile of a clogged filter shows a disparity and heterogeneity in the distribution of particle deposits.

The shear forces required to break up the deposit and enable its detachment are greater in the upper layer of the filter, where accumulation is greater than in the lower layer of the filter. In the top part of the filter, the particle load means that the optimum porosity is not far from being achieved.

The rate of the wash water must be high enough to eliminate all particles without being excessive in order to avoid media loss. It is therefore higher than the minimum fluidization velocity. A 25–30% expansion of the material is sufficient.

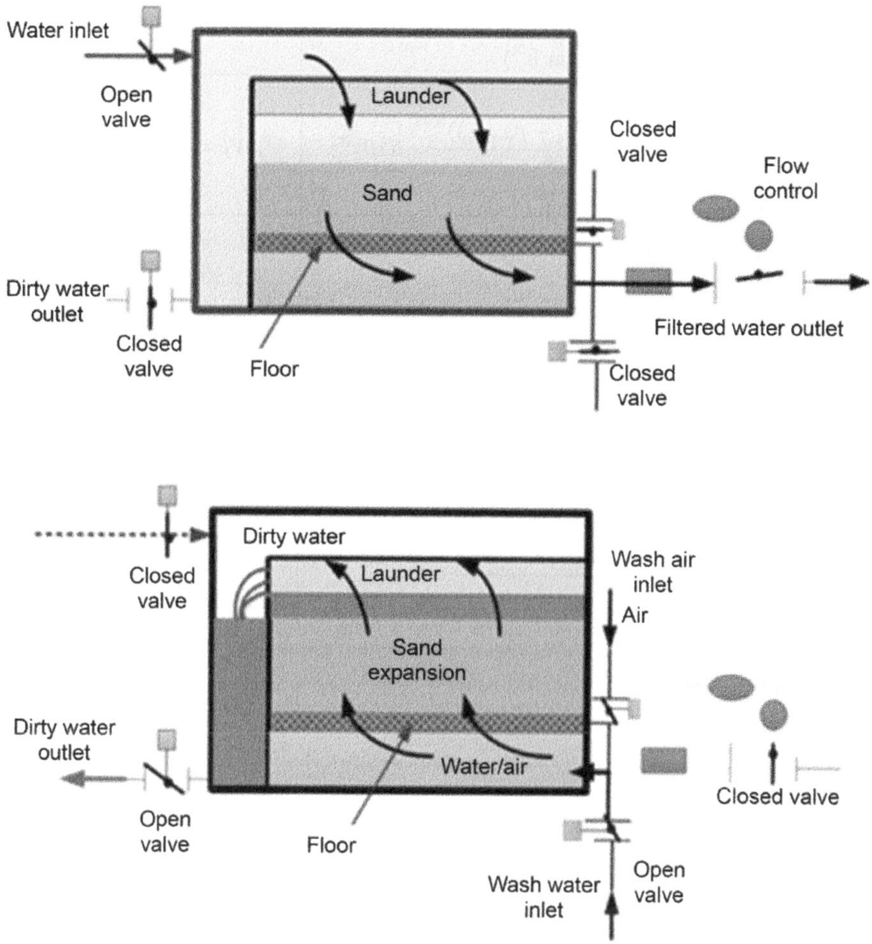

Figure 7.11. *General diagram of filter operation. For a color version of this figure, see www.iste.co.uk/gaid/reuse1.zip*

A minimum fluidization velocity followed by a washing rate complying with this level of expansion and allows the removal of the detached particles is then determined.

This washing rate is obtained with various equations shown in Table 7.8.

	Minimum fluidization velocity (V_{mf}) ($m \cdot h^{-1}$)	**Backwashing rate (V_l) ($m \cdot h^{-1}$)**
Leva	$Vmf = \frac{22.12(\rho_{water}/(\rho_s - \rho_{water}))^{0.94}}{\mu,^{0.88}} d10^{1.82}$	$Vl = Vmf\left(\frac{\varepsilon exp}{\varepsilon}\right)^n$
Ergun	$Vmf = -b + \frac{\sqrt{\Delta}}{2a} 3{,}600$	$Vl = Vmf\left(\frac{1-\varepsilon}{\varepsilon^3}\right)\left(\frac{\varepsilon exp^3}{1-\varepsilon exp}\right)$
Kozeny	$Vmf = \left(\frac{1{,}000\ (\rho_s - \rho_{water}).g.\varepsilon^3}{(hK.\mu.0.001)\ (1-\varepsilon)(\frac{6}{0.001.\psi.d10})^2}\right) 3{,}600$	$Vl = Vmf\left(\frac{1-\varepsilon}{\varepsilon^3}\right)\left(\frac{\varepsilon exp^3}{1-\varepsilon exp}\right)$
When and Yu	$Vmf = \frac{\mu}{\rho_{water} d10}(33.7^2 + 0.0408\ Ga)^{0.5} - \frac{33.7\mu}{\rho d10}$ $Ga = \frac{d^3\rho(\rho s - \rho)g}{\mu^2}$	$Vl = Vmf\left(\frac{1-\varepsilon}{\varepsilon^3}\right)\left(\frac{\varepsilon exp^3}{1-\varepsilon exp}\right)$

Table 7.8. *Equations to calculate the minimum fluidization velocity and backwashing rate*

With:

- ρ_{water}: water density ($g \cdot cm^{-3}$) for a given temperature (°C);
- ρ_s: material density ($g \cdot cm^{-3}$);
- μ: dynamic water viscosity (cPa) for a given temperature (°C);
- Ψ: media shape factor;
- d_{10}: media ES (mm);
- hK: Kozeny constant;
- ε: media porosity;
- v_{set}: particle settlement velocity ($m \cdot s^{-1}$);
- v_l: backwashing rate ($m \cdot s^{-1}$).

As the filter media volume per unit area is constant, we then have:

$$\frac{Hexp}{H} = \frac{1-\mathcal{E}}{1-\mathcal{E}exp} = \frac{1-\mathcal{E}}{1-\left(\frac{vl}{vexp}\right)^{0.22}}$$

where ε_{exp} is the expanded media porosity:

$$\mathcal{E}exp = \left(\frac{vl}{vdec}\right)^{0.22}$$

and:

$$v_l = v_{set}\, \mathcal{E}_{exp}^{\ 4.5}$$

with:

– H: media height (m);

– H_{exp}: media l height in expansion (m), $Hexp = H\ (1 + \frac{expansion}{100})$;

– Re: Reynolds number, Re = √ [33.72 + 0.0408.Ga] – 33.7;

– Ga: Galilei number;

– parameters a, b and c:

$$a = \frac{36{,}000\ hB\ \rho e}{\Psi\ d10/1{,}000}$$

$$b = \frac{36\ kH(0.001\mu)(1-\mathcal{E})}{(\Psi\frac{d10}{1{,}000})^2}$$

$$c = -1{,}000(\rho s - \rho e)9.81.\mathcal{E}^3$$

– pressure drop required to expand the filter:

$$h = Hexp\ (1 - \mathcal{E}exp)\frac{(\rho_S - \rho_{water})}{\rho_{water}}$$

H_{exp} is the expansion height, $\mathcal{E}_{exp}$ is the expanded filter porosity, and ρ_s and ρ_{water} are, respectively, the densities of the filter media and water.

Figure 7.12 shows backwashing rates for anthracite and pumice depending on expansion.

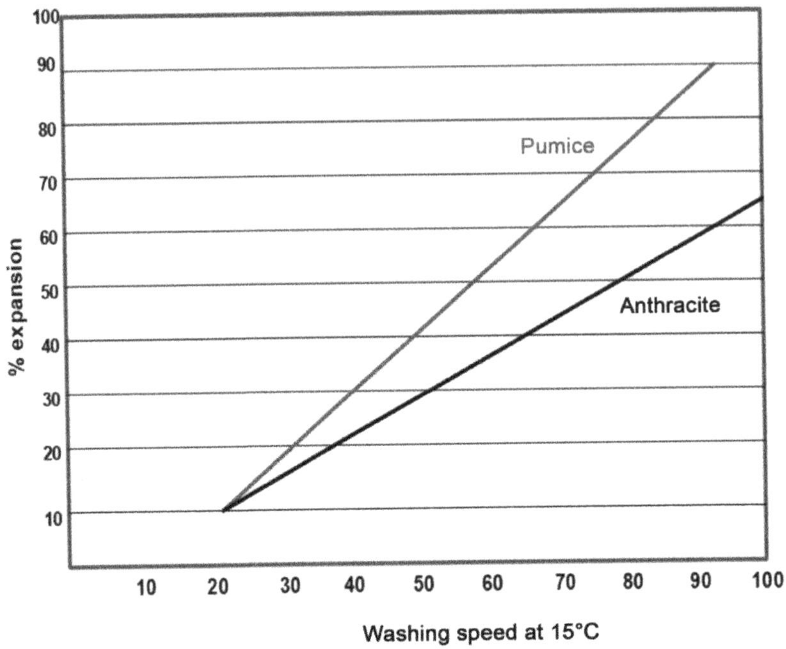

Figure 7.12. *Expansion percentage versus backwashing rate for anthracite and pumice at 15°C*

The minimum fluidization velocity, V_{mf}, is the required rate for starting the fluidization of materials. The calculation of the minimum fluidization velocity is important for determining the backwashing flow conditions. For 25–30% expansion, the diameter d_{10} is sufficient, as a diameter d_{90} involves the largest media and leads to complete fluidization of the filter.

7.6.3.2. *Implementation backwashing conditions*

The implementation of the backwashing conditions consists of three phases: air, air + water and water only.

7.6.3.2.1. Gravity filters

The water level is first lowered to the height of the media. This saves all the water above the filter media as filtered water.

First phase: air washing

During this phase, the air injection allows the filter to be loosened, which has gradually compacted over the course of the filtration cycle. It also enables breaking down the particles retained between the interstices of the material. In addition, the air breaks the surface crust on which the largest deposits of particles are found. The air injection system can be a pipe completely separated from the washing water system or be performed through nozzles distributing both air and water. If air is introduced directly under the gravel, arrangements should be made to prevent the air from moving the fine gravel, especially air and water during the air/water phase. As such, the nozzle system is more efficient and less problematic.

Air is never used during overflow (backwashing with water alone), as this results in material suspension and loss of material with dirty water. Cleaning by air scouring at the base of the filter is carried out for 1–2 min at 50 $Nm^3 \cdot h^{-1} \cdot m^{-2}$. It is necessary to ensure visually the correct distribution of air over the entire filter to detect any anomalies (broken nozzles, faulty seal).

Second phase: washing with air and water

Backwashing with water alone, without injecting air, is inefficient due to limited particle collisions and detachments as shear forces are not sufficient to break the particle/material bonds. During this air/water phase, a slow expansion of the material grains occurs, caused by the simultaneous flow of air and water, which causes abrasion between the grains. This abrasion and high water interstice speeds result in effective backwashing.

Water is sent at low flow rate simultaneously with air for about 8 min to dislodge particles between the material interstices and to suspend them in the area above the filter. The adhesion forces between the deposited particles and the filter grains can become so high that washing conditions must be adapted for particles to detach. If not, the particles and material grains can agglomerate to form sludge balls, which are very resistant to washing.

During this phase, the shear forces are high, and the rising water allows particles to be transported out of the interstices. Under these conditions, the porosity of the expanded material increases. The resulting grain movement promotes collisions and attrition between grains. This results in particles detaching, which are then driven by the water rising to the surface.

Air/water washing does not produce sludge aggregates as there is no bed expansion and the surface crusts are broken by air.

During air injection, the wash water flow rate can vary in a very wide range between 8 and 20 $m^3 \cdot h^{-1} \cdot m^{-2}$, depending on the material height. It is lower than the minimum fluidization velocity. Air flow is maintained between 50 and 60 $Nm^3 \cdot h^{-1} \cdot m^{-2}$.

The time allotted for this second phase is calculated so that the dirty water reaches the filter surface level without overflowing. Beyond this, there is a risk of material loss as part of it is suspended.

Third phase: water rinsing

The third phase involves rinsing the filter with water alone. Sufficient water flow is used to achieve expansion between 25% and 30%. The sludge separates from the surface of sand particles as they move in the colliding water flow.

The water flow varies between 20 and 60 $m \cdot h^{-1}$, depending on the media size, temperature and type of filter (single media and dual media). This operation takes 8–15 min. Rinsing continues until marbling appears on the filter surface or clear rinse water is obtained. Expansion of 25–30% is recommended, as too high an expansion results in increased grain spacing and reduced collision frequencies. The particle detachment efficiency is very low.

This type of backwashing is effective if the surface washing speed is greater than the minimum fluidization velocity of the material. Otherwise, mud balls can form, and pressure drops could develop faster.

It should be noted that the backwashing rate is different between summer and winter, as it depends on the water viscosity. To maintain the same expansion, a lower flow rate is normally used during cold temperatures compared to the temperature used in warm periods.

Depending on the design of the filter, washing may sometimes be combined with surface washing, which consists of injecting additional water, which helps to evacuate flocs, especially when the height of water above the filter is significant (>1.4 m). This prevents re-flocculation of sludge, which grows larger, becomes heavier and cannot be removed at the conventional backwashing rate. The volume of rinse water varies from 2 to 5% of the treated flow rate.

After the fluidization has stopped, materials settle according to Stokes' law (laminar settling).

In dual-media filters, backwashing sequences are identical in terms of the operating protocol. During the air/water phase, the air flow is kept for 1–2 min. The associated water flow is rather between 12 and 20 $m \cdot h^{-1}$ for 8 min, depending on the type of media, its ES and the height of the media.

Several precautions are recommended to avoid the loss of sand or anthracite during the air/water phase. Water and air flow rates vary depending on the size of the filter media. Generally, dirty water will reach the top of the filter walls, and the vertical distance between the surface of the anthracite and the wash water overflow varies from 0.6 to 1.4 m depending on the filters.

During the rinsing phase (water alone), the backwashing rate is greatly increased to 50–60 $m \cdot h^{-1}$ in order to partially fluidize the different materials and then restore stratification by gravity separation. In dual-media filters, it is not recommended to have a mixture of media layers as the finer media that is in the lower layer plays the role of turbidity refining.

The equation for calculating the difference between the apparent densities of two media is:

$$\rho_{\text{filter}} = (1 - \varepsilon 1)(\rho_s 1 - \rho_{\text{water}}) - (1 - \varepsilon 2)(\rho s 2 - \rho_{\text{water}})$$

where ρ_{filter} is the apparent density of the filter and ρ_s is the grain density of the two media.

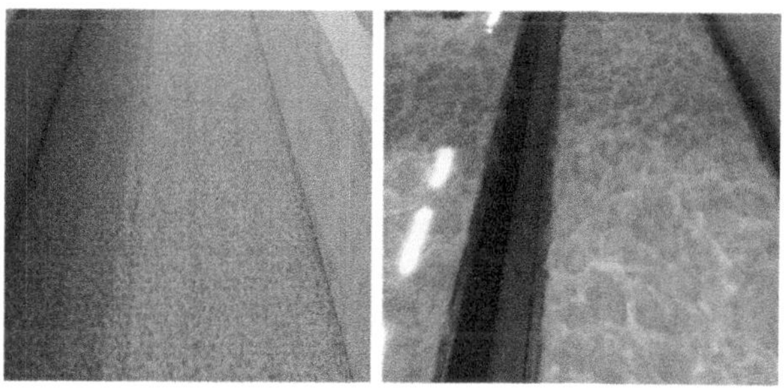

Figure 7.13. *First decompaction phase with air and second (air/water) washing phase. For a color version of this figure, see www.iste.co.uk/gaid/reuse1.zip*

$\rho_{\text{filter}} = 0$ means that both materials are completely mixed. A positive or negative value of this parameter means that there is stratification between both materials.

Index 1 is associated with the largest media (anthracite) and index 2 with the finest media (sand). As material 2 has a higher density than that of material 1, material 1 tends to fall below material 2 for low wash water flow rates. For a sufficiently high flow rate, the coarser material 1 is above the thinner material 2. This is the case between anthracite or pumice and sand.

Fourth phase: filter back in production

When the required turbidity of filtered water is low (<2 NTU), it is recommended that the first filtered water be returned to the treatment plant during the first 10–30 min when filters are back in production.

When the first filtered water cannot be discharged in sewer, it is recommended to use a slow filter restart in order to limit the release of particles into the filtered water.

Care should be taken not to create filtration over speeds when the filter, which has just been washed, is put back in service, and generally not to create hydraulic shocks on the whole filter battery when the washed filter is put back in service.

Table 7.9 summarizes the different washing conditions for some single-layer and two-layer filters.

Parameters	**Sand filter H: 1 m ES: 0.85 mm**	**HR sand filter H: 2 m, ES: 1.35 mm**	**Anthracite/sand dual-media filter H: (0.8 + 0.8) m ES: 1.4 mm ES: 0.85–1 mm**	**Pumice stone/sand dual-media filter H: (0.8 + 0.8) m ES: 1.4 mm ES: 0.85–1 mm**
Air	50 $Nm^3 \cdot m^{-2} \cdot h^{-1}$ (1 min)	55–60 $Nm^3 \cdot m^{-2} \cdot h^{-1}$ (1 min)	50 $Nm^3 \cdot m^{-2} \cdot h^{-1}$ (1 min)	50 $Nm^3 \cdot m^{-2} \cdot h^{-1}$ (1 min)
Air + water	50 $Nm^3 \cdot m^{-2} \cdot h^{-1}$ for air 8 $m^3 \cdot m^{-2} \cdot h^{-1}$ for water (8 min)	55–60 $Nm^3 \cdot m^{-2} \cdot h^{-1}$ for air 15 $m^3 \cdot m^{-2} \cdot h^{-1}$ for water (8 min)	50 $Nm^3 \cdot m^{-2} \cdot h^{-1}$ for air 12 $m^3 \cdot m^{-2} \cdot h^{-1}$ for water (8 min)	50 $Nm^3 \cdot m^{-2} \cdot h^{-1}$ for air 12 $m^3 \cdot m^{-2} \cdot h^{-1}$ for water (8 min)
Rinsing	25 $m^3 \cdot m^{-2} \cdot h^{-1}$ for water (7–10 min)	60 $m^3 \cdot m^{-2} \cdot h^{-1}$ for water (10–15 min)	50-55 $m^3 \cdot m^{-2} \cdot h^{-1}$ for water (7–10 min)	35 $m^3 \cdot m^{-2} \cdot h^{-1}$ for water (7–10 min)

Table 7.9. *Backwashing conditions for single-media and dual-media filters*

7.6.3.2.2. Media expansion during backwashing

When using a backwash with about 20–30% fluidization, this expansion depends on the type of the media and water parameters. The parameters related to the media are the diameter as well as its shape and density. The parameters related to water are viscosity and density. The estimation of expansion is important, for example, to know how high the media will rise and to prevent it from reaching the filter channels.

As such, a modified Reynolds number is calculated:

$$Re\,mod = \frac{V\,\varepsilon\,exp\,\rho}{\varepsilon\,exp\,Sv(1-\varepsilon exp)\mu} = \frac{V\rho}{Sv(1-\varepsilon exp)\mu}$$

where Sv is the media specific surface area (6/d for spheres and 6/ψ d for non-spherical grains).

With A as dimensionless number related to porosity and specific surface area:

$$A = \frac{\varepsilon exp^3 \rho_{water}(\rho_S - \rho_{water})g}{(1-\varepsilon exp)^2 Sv^3 \mu^2}$$

Log A is correlated with log Re mod to obtain:

$$\log A = 0.56543 + 1.09348\,(\log Re\,mod) + 0.17971\,(\log Re\,mod)^2 - 0.00392\,(\log Re\,mod)^4 - 1.5(\log \psi)^2$$

By replacing the equation of A and Re mod including porosity, the expanded value of ε is deduced.

Namely:

$$\log\left(\frac{\varepsilon exp^3 \rho_{water}(\rho_S - \rho)g}{(1-\varepsilon exp)^2 Sv^3 \mu^2}\right) = 0.56543 + 1.09348\left(\log\left(\frac{V\rho}{Sv(1-\varepsilon exp)\mu}\right)\right) + 0.17971\left(\log\left(\frac{V\rho}{Sv(1-\varepsilon exp)\mu}\right)\right)^2 - 0.00392\left(\log\left(\frac{V\rho}{Sv(1-\varepsilon exp)\mu}\right)\right)4 - 1.5(\log \psi)^2$$

As a result, the expansion height (H_{exp}) is deduced from the equation:

$$\frac{Hexp}{H} = \frac{1-\varepsilon}{1-\varepsilon exp}$$

Table 7.10 summarizes the various parameters involved in the size and operation of tertiary filters.

Filtration parameters	Comments
Filtered water quality *Filter feed water characteristics* SS concentration Turbidity	Acceptable maximum pressure drop Addition of reagents Backwashing conditions Set by local regulation Affect operating and conditions
Media characteristics	Affect particle elimination and pressure drop
Effective size d_{10} Uniformity coefficient Shape, density, type factors	
Filter characteristics	Height affects the initial pressure drop and the cycle duration Porosity affects the matter quantity that can be retained
Filtration height(s)	
Porosity	Affect filter performances and cycle duration
Stratification Filtration rate	Affects cycle duration Affect equipment size and material classification after washing

Table 7.10. *Main parameters involved in filter size*

7.7. Filtration technologies

In general, filters have the following common characteristics:

– filtration in the downward direction and backwashing in the upward direction (backwash);

– a false bottom composed of a filter floor made of slabs and nozzles or Leopold type filter bottoms or equivalent;

– a 100 mm layer of gravel 4–7 mm covering the heads of nozzles for filter floors;

– one or more layers of filter materials;

– a constant water level during the filtration phase, maintained by downstream control;

– a two-level channel (between two filters), the lower part of which is designed to collect filtered water, on the one hand, and to distribute dirty water, on the other hand; the upper part is mainly used for collecting dirty washing water;

– washing equipment adapted to the filter characteristics.

According to the work technology, we distinguish:

a) open filters, where water is gravity fed at atmospheric pressure, are generally the most used;

b) pressure filters, where water passes under pressure through the filter material;

c) mobile filters.

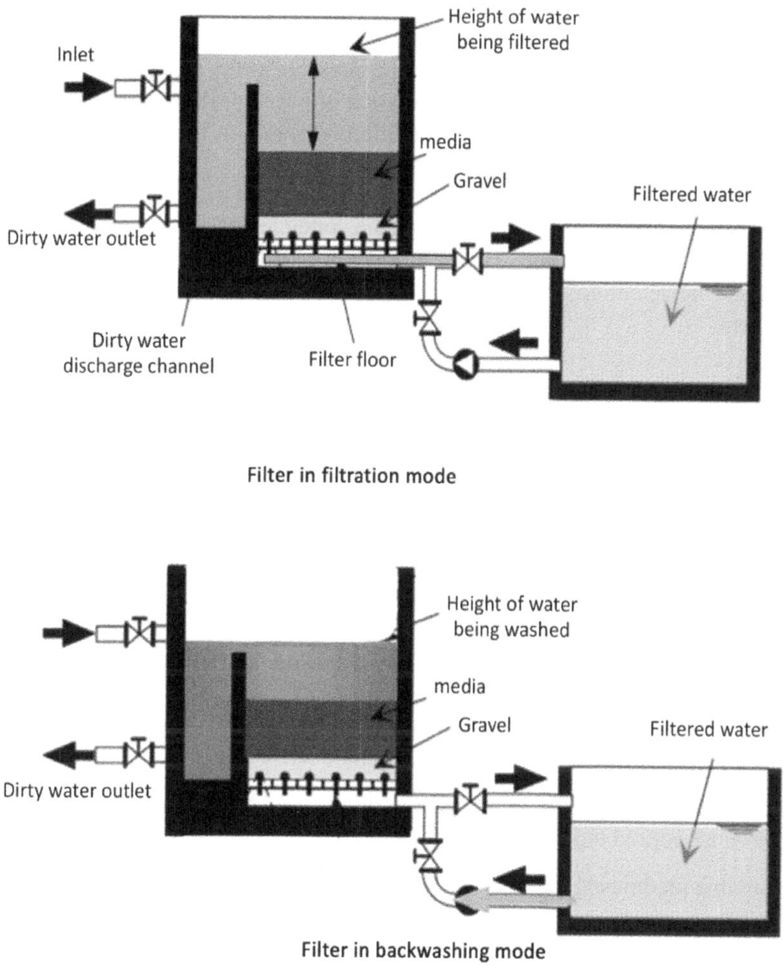

Figure 7.14. *Single-media filter in filtration mode and backwashing mode. For a color version of this figure, see www.iste.co.uk/gaid/reuse1.zip*

Characteristics regarding sand, media height, filtration rate, retention capacity and backwashing conditions of rapid or HR sand filters differ. The media is laid out on a nozzle floor, or Leopold bottoms or equivalent. The design of the nozzle floor includes a surface made of reinforced concrete slabs that are equipped with nozzles. These have dimensions that depend on the sand grain size and the thickness of the slab.

The Leopold system consists of a series of PVC blocks that form the floor surface and are equipped with mesh whose sizes are sufficient to hold everything while letting filtered water pass. When the pressure drop reaches its permissible limit due to filter clogging by the accumulation of particles, the filter is backwashed with the filtered water. The washing conditions depend on the filter material grain size.

HR filters, whether single or dual media, operate at rates two to three times higher than conventional filters. Because of the deep diffusion of particles within the filter mass, the retention capacities of these filters are much higher than those of rapid filters.

The distinction between rapid filters and HR filters also comes from the mechanism to eliminate particles: rapid filters retain particles on the surface of the filter material while HR filters retain them on a large part of the filter mass.

7.7.1. *Conventional single-layer filters*

These filters are the most common, thanks to their simplicity and low investment cost. They are characterized by a water height in operation of 0.6–1.3 m, the advantages of which are a fast discharge of wash water and low water loss.

These are open gravity flow filters whose rectangular-shaped casing is most often built in reinforced concrete. They are characterized by:

– a single filter layer of homogeneous grain size resting on a nozzle draining floor or on a Leopold type floor, or equivalent;

– variable media height;

– a three-phase backwash (air, air/water and water-only rinse).

The media grain size depends on the field of application, the rate and the filtration height: the floor of the filter layer, made of reinforced concrete slabs, is equipped with nozzles at a density of 45–50 nozzles/m^2 of floor slab. These nozzles are buried in a layer of gravel. The slabs are supported by beams. Holes are provided in the beams (using rigid PVC tubes embedded in concrete) to facilitate the redistribution of the wash air along the filter length.

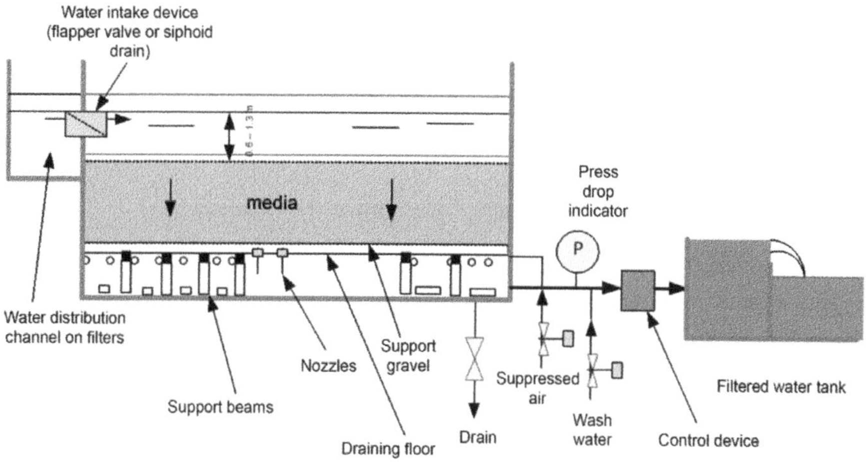

Figure 7.15. *Conventional filter*

The backwash cycle consists of:

– lowering the water level to the material level, followed by air injection (50 $Nm^3 \cdot m^{-2} \cdot h^{-1}$) for 1–2 min;

– the air/water phase is then implemented with filtered water;

– the air/water phase is carried out with an injection of air (50 $Nm^3 \cdot m^{-2} \cdot h^{-1}$) and filtered water in the opposite direction at a flow of 8–15 $m^3 \cdot m^{-2} \cdot h^{-1}$. The time allotted to this phase is 8 min but can be extended to 12 min, if necessary. This phase must not allow water to overflow over the filter level surface, which could result in material loss;

– the rinsing phase is carried out with filtered water alone (25–60 $m^3 \cdot m^{-2} \cdot h^{-1}$) depending on the type of material and operating conditions. The time allotted for this phase is 8 min but can be extended to 15 min, if necessary. Usually, it is recommended to maintain the flow of backwash water until the drained water is clear;

– the discharge of wash water is made by discharge on one or both sides, depending on the width of the cell, as shown in Figure 7.16.

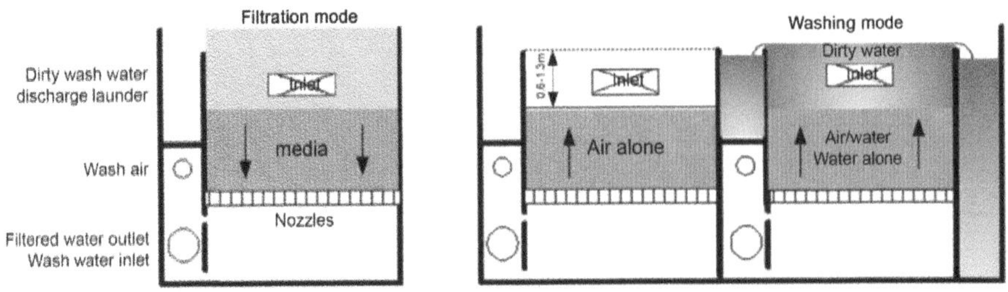

Figure 7.16. *Filtraflo F washing operating mode*

The volume of the dirty water receiving cover is calculated on the basis of two to three volumes of media. The first filtered water after backwashing is most often directed to the sewer or sometimes recirculated (according to local regulations). There is a pipe with vents on the wash water collector after the first water outlet valve (one valve for all filters) of the same diameter as the filtered water outlet valve of each filter.

7.7.2. *High-rate filters*

They are a variation of conventional filters. These are open gravity flow filters whose rectangular-shaped casing is most often made of reinforced concrete. They are characterized by:

– a single filter layer of homogeneous grain size resting on a nozzle draining floor or on a Leopold type floor or equivalent;

– media height variable between 1.5 and 2.2 m;

– ES of 1.35 mm and CU < 1.4;

– three-phase backwash (air, air/water and water alone) (see Table 7.9).

The principle of HR filtration is to increase the filtration height and the filter material grain size in order to increase the storage capacity of SS in the filters. Indeed, the operating principle is a diffusion of flocs in depth in the filter mass supported by a HR filtration. The pressure drop observed in the filter is linear.

Increasing the media height means that a higher pore volume is available for particle retention, which in turn improves the overall performance of the filter. As such, the recommended H/d_{10} ratio must be greater than 1,300. However, the increase in sand height is not significant if it is not accompanied by a high hydraulic load, otherwise most of the SS will simply accumulate in the upper part of the sand bed. Increased water load improves filtration cycles, as a higher maximum available pressure loss is available in this case.

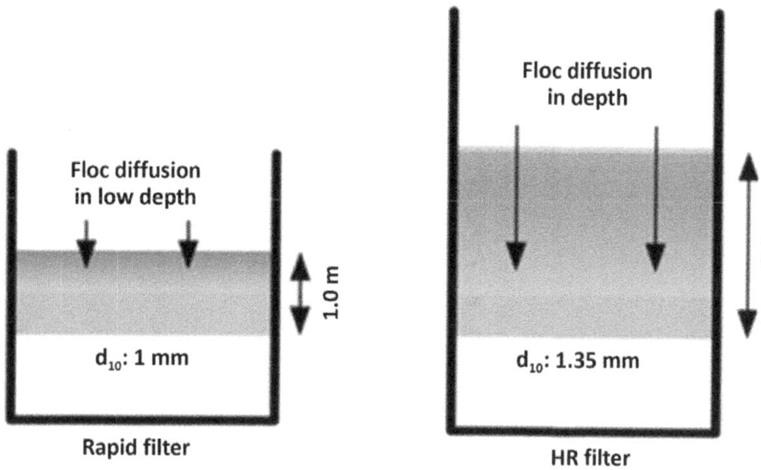

Figure 7.17. *Diagram of floc diffusion in a rapid filter and a HR filter*

The HR filters operate at rates of 14–16 m in N and 20 m·h^{-1} in N-1. As a result, the floor area and number of filters are reduced. The retention capacity is expressed in kg.m^{-3} as the high filtration rate causes particles to diffuse at least 1.5 m deep. It is around 3–4 kg SS·m^{-3} of media volume.

The water height above the filter mass is calculated taking into account:

– the minimum water height to avoid too significant low-pressures in the filter mass;

– the minimum permissible water height for washing to ensure proper discharge of accumulated SS during an operating cycle.

The minimum water height in filtration shall not be less than 0.8 m above the material level. The permissible wash water height for good SS discharge is 1.4 m. The available pressure drop on the filter mass (clean filter pressure drop + pressure drop during the filtration cycle) is then calculated at the maximum flow rate, on N-1 filters taking into account also the pressure loss brought by the nozzle floor, as well as that of the valves and filtered water pipes.

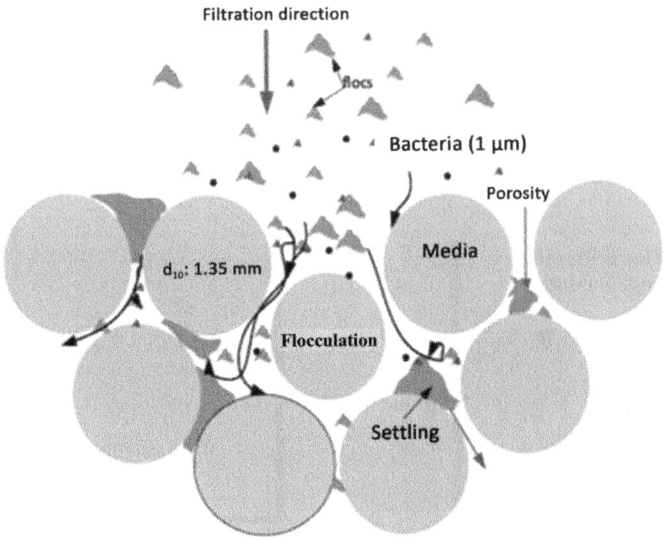

Figure 7.18. *Particle attachment mechanisms on HR filters*

In the case of an individual restitution basin, the maximum available pressure loss on filter mass is about:

– 2.6 mCE for a media height of 2 m;

– 2 mCE for a media height of 1.5 m.

In the case of a significant change in flow, during the low flow rate, the HR filter can behave like a conventional filter at 10 m·h^{-1} and therefore clog on the surface, which leads to a faster fouling of the filter and the need to wash more often; the notion of clogging degree must therefore be taken into account, according to the following formula:

$$d°\,(\%) = 100\frac{\Delta P(t)}{\Delta Pmax}\left(\frac{Vmax}{V(t)}\right)^{\alpha}$$

with:

– Vmax: maximum filtration rate (m·h^{-1});

– V(t): filtration rate at time (m·h^{-1});

– Δpmax: permissible maximum pressure drop of the media (mCE);

– Δp(t): media pressure drop right now (mCE).

This degree of clogging enables to integrate the physical fouling of filters with some series washes when returning to HR flows. The α coefficient can be configured at commissioning with a value between 0.5 and 1.0. The wash cycle consists of:

– lowering the water level to the media level, followed by air injection (50 $Nm^3 \cdot m^{-2} \cdot h^{-1}$) for 1–2 min;

– the air/water phase is then implemented with filtered water;

– the air/water phase is carried out with an injection of air (50 $Nm^3 \cdot m^{-2} \cdot h^{-1}$) and filtered water in the opposite direction at a flow of 8–15 $m^3 \cdot m^{-2} \cdot h^{-1}$. The time allowed to this phase is 8 min but can be extended to 12 min, if necessary. This phase must not allow water to overflow over the filter level surface, which could result in material loss;

– the rinsing phase is carried out with filtered water alone (25–60 $m^3 \cdot m^{-2} \cdot h^{-1}$) depending on the type of material and operating conditions. The time allotted for this phase is 8 min but can be extended to 15 min, if necessary. Usually, it is recommended to maintain the flow of backwash water until the drained water is clear.

The inlet to the filter is composed of pipes with isolation butterfly valves and a flow diffusion optimization and anti-scour device. Indeed, given the high speeds, scour phenomena can occur. In addition, a good flow diffusion using diffusion cones must be implemented to avoid excessive turbulence.

The floor of the reinforced concrete slab filter layer is equipped with long-tailed nozzles at a density of 50–65 nozzles/m^2 of floor slab. These nozzles are buried in a layer of gravel. The slabs are supported by beams. Holes are provided in the beams (using rigid PVC tubes embedded in concrete) to facilitate the redistribution of the wash air along the length of the filter. The volume of the dirty water receiving cover is calculated on the basis of two to three volumes of material.

The first filtered water after washing is most often directed to the sewer or sometimes recirculated (according to local regulations). There is a pipe with vents on the wash water collector after the first water outlet valve (one valve for all filters), of the same diameter as the filtered water outlet valve of each filter. On the other hand, this first filtered water can also not undergo any treatment and be sent to the disinfection stage. This protocol depends on the regulations applied in each country.

HR filters have longer cycle times due to their higher retention capacity than rapid filters.

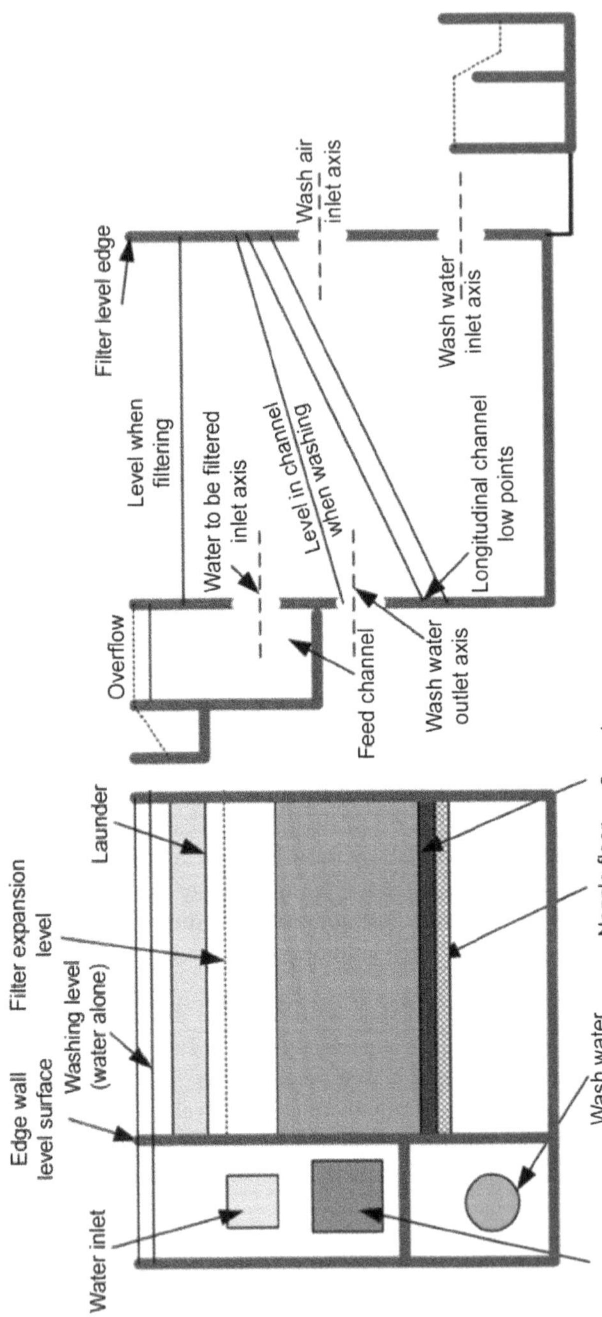

Figure 7.19. *HR filter in operation and washing*

7.7.3. *Example of tertiary filtration design on single-media filter*

7.7.3.1. *Calculations based on N-2 filters at the maximum flow rate*

Parameters	Units	Design values	Maximum authorized flow rate	Maximum daily flow rate
Daily average flow rate	$m^3 \cdot day^{-1}$	111,473	126,477	154,473
Peak flow rate	$m^3 \cdot day^{-1}$	6,134	6,947	6,947
Total number of filters	–	8		
Unit area of a filter	m^2	99.63		
Filter size	m × m	6.09 × 16.36		
Sand material height	m	2		
Sand ES	mm	1.35		
Sand UC	–	1.3–1.4		
Gravel height	m	0.1		
Gravel ES	mm	6.0		
Water height above material	m	1.20		
Water height under filter floor	m	1.3		
SS filter retention capacity	$kg \cdot m^{-3}$	2.5		
N filters rate	$m \cdot h^{-1}$	5.83	6.61	8.10
N-2 filter rate (washing period)	$m \cdot h^{-1}$	7.77	8.81	10.76
N rate in peak flow period	$m \cdot h^{-1}$	7.69	8.71	8.71
N-2 rate in peak flow period	$m \cdot h^{-1}$	10.26	11.62	11.62
SS inlet	$mg \cdot L^{-1}$	20	20	20
SS outlet (expected)	$mg \cdot L^{-1}$	5–10	5–10	5–10
Turbidity (expected)	NTU	2	2	2
Cycle duration (h) on max. daily flow rate (24-h guaranteed value)	N/N-1	37–43	33–38	27–37
Recirculated water volume (dirty wash water)	$m^3 \cdot day^{-1}$	8,502	8,502	502

Table 7.11. *Tertiary filtration design on single-media filter*

7.7.3.2. *Washing conditions*

Parameters	Units	Values
	Phase 1 (air alone)	
Decompaction air	$Nm^3 \cdot h^{-1} \cdot m^2$ × duration	60 × 2 min
	Phase 2 (air + water)	
Low water flow rate	$m^3 \cdot h^{-1} \cdot m^2$ × duration	20 × 8 min
Wash air	$Nm^3 \cdot h^{-1} \cdot m^2$ × duration	60 × 8 min
	Phase 3 (water alone)	
High water flow rate	$m^3 \cdot h^{-1} \cdot m^2$ × duration	60 × 8 min

Table 7.12. *Washing conditions of a single-media filter*

7.7.3.3. *Characteristics of washing pumps and air compressors*

Parameters	Units	Values
Number of washing pumps on standby	–	3 1
Flow per pump **HMT**	$m^3 \cdot h^{-1}$ m	3,000 8.5
Recirculation pump rate **HMT**	$m^3 \cdot h^{-1}$ m	486 19.5
Number of compressors for washing on standby	–	2 1
Flow per compressor **Differential pressure**	$m^3 \cdot h^{-1}$ mbar	6,000 650

Table 7.13. *Filter washing equipment*

7.7.4. *Dual-media filters*

Dual-media filters are open gravity flow filters whose rectangular casing is most often constructed of reinforced concrete. They are characterized by:

– a double filter layer resting on a nozzle draining floor or on a Leopold type floor or equivalent;

– a large water height above the filter materials (1.2–1.8 m), as it allows to ensure the degassing of previously aerated or ozonized water, sufficient flocculation time in case of direct filtration, longer cycle times due to higher available pressure drop;

– total material height between 1.2 and 2.0 m;

– three-phase backwash (air, air/water and water alone).

They are installed after coagulation-flocculation-settlement, regardless of the type of settler. They are also offered in direct filtration after coagulation (in-line or in a tank) and flocculation in a tank or in-line. They are also offered in the case of a settler by-pass when raw water turbidity is lower at certain periods of the year. Dual-media filters are used in the same basic configuration as rapid filters. The materials of the upper layer have a lighter density than the material of the lower layer, but a higher porosity. Sand porosity is typically 45% while anthracite porosity is 50–55%. As a result, expansion during washing and reclassification after washing of both materials are easy to implement.

For example, material combinations can be:

– anthracite/sand;

– pumice/sand.

Filtration rates depend on the total material height. They can vary between 8 and 17 $m \cdot h^{-1}$ in N. The water intake into the filters is provided from a general distribution channel by a submerged tubing equipped with an on-off isolation valve. A deflector is placed under the valve to prevent the scouring of material layers. The filtered water is collected in a duct under the launder collecting dirty water from the washing. It is then returned by an outlet tubing equipped with a control system in a receiving cover for open body of water from which it is discharged by a spillway. The filtered water collection duct is also used to provide the distribution of wash water and air. The wash cycle consists of:

– lowering the water body to the level of the top layer material followed by air injection (50 $Nm^3 \cdot m^{-2} \cdot h^{-1}$) for 1–2 min;

– the air/water phase is then implemented with filtered water. It is carried out with an injection of air (50 $Nm^3 \cdot m^{-2} \cdot h^{-1}$) and filtered water in counterflow at a flow rate of 8–15 $Nm^3 \cdot m^{-2} \cdot h^{-1}$. The time allotted to this phase is 8 min but can be extended to 12 min, if necessary. This phase must not allow water to overflow over the filter level surface, which could result in material loss;

– the rinsing phase is carried out with filtered water alone (35–60 $m^3 \cdot m^{-2} \cdot h^{-1}$) depending on the type of material and operating conditions. The time allotted for this phase is 8 min but can be extended to 15 min, if necessary. Usually, it is recommended to maintain the flow of backwash water until the drained water is clear.

The backwash will be adapted to the media used (depending on the expansion rate of the anthracite).

The support floor of the filter layers is made up of reinforced concrete slabs, equipped with long-tailed nozzles with a density of 50 units/m^2 of floor slab. The nozzle heads are embedded in a layer of gravel on which the layers of materials rest. The floor slabs are supported by reinforced concrete beams.

The washing fluids (air and water in counterflow) are distributed under the floor from a single duct under the dirty water collection launder. The distribution system is composed of rectangular holes level with the raft for water and under-floor tubes for air injection.

As with other filtration technologies, the wash cycle is initiated on a pressure drop level, on an output turbidity, on a clock or on the operator's decision.

Figure 7.20 shows a single filter or two-filter configuration, with filtration and washing phases.

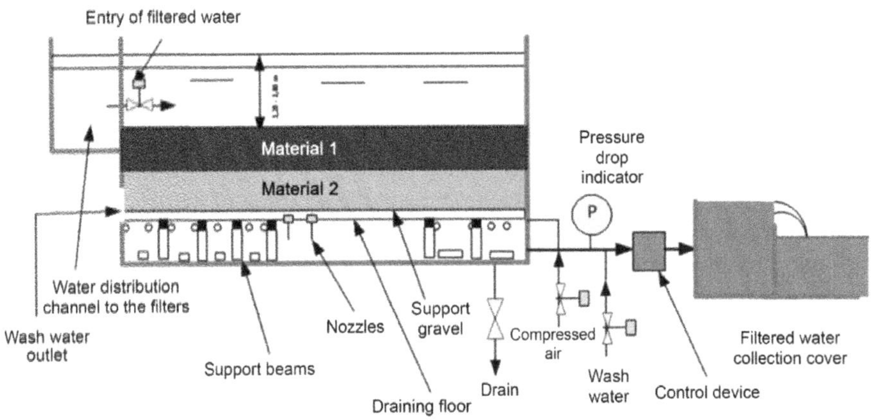

Figure 7.20. *Description of a dual-media filter*

7.7.4.1. *Design of a dual-media tertiary filter: example of a WWTP (Abu Dhabi, Middle East)*

The average SS content of the clarified water is set at 15 mg·L^{-1} and can reach a maximum concentration of 20 mg·L^{-1}.

The turbidity guarantee on tertiary treatment outlet water is ≤2 NTU, given on 24-h composite samples. Based on the assumption of the empirical relationship that 1 NTU = 1.5 mg·L^{-1} of SS, the volumes of water recovered during the wash cycles and intended to be recycled are calculated with this formula. The total pressure drop through the filter (water level in the filter – water level in the outlet control basin) is set at 3.00 m of water column.

Other guarantees are presented in Table 7.14.

Parameters	Units	Guarantee on a 24 h sample
pH	–	6.5–8.5
COD	$mg \cdot L^{-1}$	–
BOD5	$mg \cdot L^{-1}$	≤10
SS	$mg \cdot L^{-1}$	≤10
NH_4-N	$mg \cdot L^{-1}$	≤2
Dissolved oxygen	$mg \cdot L^{-1}$	≥3
Turbidity	NTU	≤2
E. coli	CFU	≤100/100 mL
Free chlorine residual	$mg \cdot L^{-1}$	≥1

Table 7.14. *Guarantees of treated wastewater quality for irrigation application*

In order to satisfy a turbidity of less than 2 NTU, the choice was made for a dual-media pumice/sand filtration.

The distribution of the flow of wastewater to the filters will be ensured by the implementation of:

– a calibrated orifice plate upstream of the main distribution channel of each filter line allowing 10 cm of pressure drop;

– a calibrated orifice at the inlet of each filter allowing 25.4 cm of pressure drop;

– a calibrated orifice downstream of the inlet of each filter and upstream of the internal distribution channels (concrete) allowing 10 cm of pressure drop;

– a spillway ensuring a load break between the filter and inlet channel;

– a pneumatically operated intake valve to reduce the intensity of surface sweeping during backwashing.

The water level in the filters is maintained by control valves and level sensors.

Description	Values	
Characteristics of dual-media filters		
Contractual water flow rate including filter wash water recirculation	317,214 $m^3 \cdot j^{-1}$	
Contractual outflow rate	301,214 $m^3 \cdot j^{-1}$	
Peak water inflow rate	16,700 $m^3 \cdot h^{-1}$	
Peak water outflow rate	16,000 $m^3 \cdot h^{-1}$	
Total number of filters	14	
Filter unit area	124.5 m^2	
Filter unit size	6.09 m × 20.44 m	
Pumice height	1.4 m	
Pumice ES	1.2–1.4 mm	
Pumice UC	1.5	
Sand height	0.6 m	
Sand ES	1.0 mm	
Sand UC	<1.5	
Water height over upper media	1.20 m	
Water height under filter floor	1.35 m	
Filter retention capacity	2.4 $kg \cdot m^{-3}$	
Filter washing conditions		
Air	50 $Nm \cdot h^{-1}$ × 2 min	
Low water flow rate (with air)	20 $m \cdot h^{-1}$ × 8 min	
High water flow rate rinsing	45 $m \cdot h^{-1}$ × 8 min	
Water volume per wash	1,080 m^3	
Wash water basin volume	2,200 m^3	
Number of washes per day	Each filter is washed at least once a day	
Wash water daily average volume	15,106 $m^3 \cdot j^{-1}$	
Filtration rate	Contract	Maximum flow rate
Rate (N)	7.58 $m \cdot h^{-1}$	9.58 $m \cdot h^{-1}$
Rate (N-1)	8.17 $m \cdot h^{-1}$	10.32 $m \cdot h^{-1}$
Cycle duration between each wash for N rate and inlet average SS	24 h	
Filter washing conditions		
Number of washing pumps	2 operating + 1 on standby	
Flow rate per washing pump Total pressure loss	2,950 $m^3 \cdot h^{-1}$ 4 m (nominal); 8 m (maximum)	
Number of wash water recirculation pumps	1 operating + 1 on standby	
Wash water recirculation/pump	750 $m^3 \cdot h^{-1}$	
Number of blowers for washing	1 operating + 1 on standby	
Flow rate per blower Differential pressure	6,500 $m^3 \cdot h^{-1}$ 650 mbar	

Table 7.15. *Tertiary two-layer filtration design*

The design of tertiary filtration is based on the following elements: the average filtration rate in N filters is set at 9.58 $m \cdot h^{-1}$ in peak hydraulic conditions for N filters in operation, 10.32 $m \cdot h^{-1}$ in N-1 conditions and 11.18 $m \cdot h^{-1}$ in N-2 conditions. For this purpose, 14 filters will be provided. The different heights of media are as follows:

– 1,400 mm of pumice;

– 600 mm of sand;

– 100 mm of gravel.

The nozzles are integrated into the load-bearing floor. The nozzles are specially designed for backwashing with air and water. The backwashing is intended to wash the SS that fill the medium with a fluid flow upstream.

The backwash process of the filters will be performed in the following cycle:

1) Wastewater discharge: the filtered water outlet valve will be closed, and the water level is lowered after opening the discharge valve.

2) Air decompaction: the air is injected for 2 min through the air inlet valves at a speed of 50 $Nm \cdot h^{-1}$ using two air blowers.

3) Air–water combination: the speeds of air and water are 50 $Nm^3 \cdot m^{-2} \cdot h^{-1}$ and 20 $m^3 \cdot m^{-2} \cdot h^{-1}$, respectively. This will last between 8 and 10 min depending on the level of fouling of the filter.

4) Rinsing: the water taken from the backwash water tank will be pumped to the filters and injected at a speed of 45 $m^3 \cdot m^{-2} \cdot h^{-1}$.

	Inflow rate $m^3.j^{-1}$	**Inflow rate $m^3.h^{-1}$**	**N filtration rate**	**N-1 filtration rate**	**N-2 filtration rate**	**Inlet SS $mg \cdot L^{-1}$**	**Outlet SS $mg \cdot L^{-1}$**	**Outlet turbidity NTU**	**Guaranteed turbidity 24-h average NTU**
Daily average flow rate	317,214	13,217	7.58	8.17	8.85	15	2	1.3	2
Daily maximum flow rate	364,796	15,200	8.72	9.39	10.17	15	2	1.3	2
Peak hourly flow rate	–	16,700	9.58	10.32	11.18	20	5	3.3	–

Table 7.16. *Synthesis of design values of the tertiary dual-media filtration*

This approximately 8-min rinsing process ensures that the mass of solids accumulated in the filter will be removed.

The results of SS and turbidity analyses show that the guarantee is respected on a 24-h average sample. SS are between 2 and 3 mg·L^{-1} while turbidity is practically always less than 2 NTU.

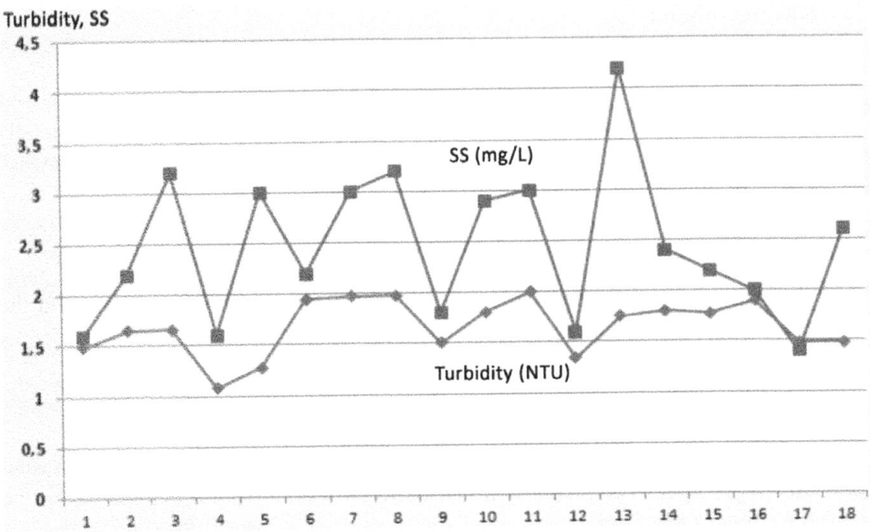

Figure 7.21. *Turbidity analysis results and SS at the outlet of tertiary filtration*

7.7.5. *Filters under pressure*

These are closed filters operating under pressure, whose construction enclosure, usually metal, is made up of a cylindrical ferrule arranged vertically and closed by two pressed bottoms.

They are characterized by:

– a filter layer with one or more materials;

– a water backwash according to the three-phase protocol;

– filtration rate: 6–18 m·h^{-1};

– filter layer(s): VF filters can contain one, two or three materials according to the same principle as described for single-layer or two-layer filters.

Generally, closed filters flow under a pressure that corresponds at least to the upper level of the filter layer. The pressure drop of a closed filter is controlled by measuring the difference in pressure between the inlet piping for water to be filtered and the outlet piping for filtered water (upstream and downstream pressure gauges). The pressure drop of a series of filters can be controlled by a single set of pressure gauges (one on the upstream collector and the other on the downstream collector). The filters of a same series usually present different states of clogging; the total flow treated is distributed unevenly between each of them, as clean filters are more used than already soiled filters.

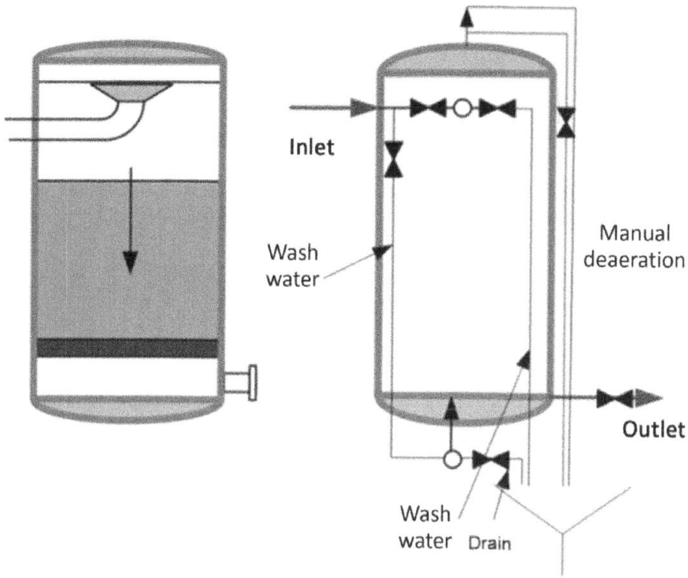

Figure 7.22. *Example of a Veolia filter under pressure*

To ensure that filter flow rates remain equal at all times, we can consider:

– giving them flow control devices;

– completing a full equidistribution of the total flow rate on units in service;

– washing them in succession, which is only possible if a large supply of wash water is available.

Figure 7.23. *Filters under pressure. For a color version of this figure, see www.iste.co.uk/gaid/reuse1.zip*

7.7.6. *Performances of gravity filters*

Filter type	Reduction log					
	Giardia	*Cryptosporidium*	Bacteria	Phage MS2	Helminth eggs	Filtered water turbidity (NTU)
Rapid filter without reagents	1.0	1.0	1.5–2	1.0	1.0	<0.5
with reagents	2–2.5	2–2.5	3	1.5–2	2.0	<0.2
Dual-media filter	2.5–3.0	2.5–3.0	3	2.5	2.0	<0.2
HR filter	2–2.5	2–2.5	2.0	1.0–1.5	1–1.5	<0.3

Table 7.17. *Log of microorganism reduction during filtration under filtered water turbidity conditions*

Rapid sand filtration eliminates approximately one to two logarithmic units of fecal coliforms, pathogenic bacteria (*Salmonella* and *Pseudomonas aeruginosa*) and enteroviruses, 90% of protozoan cysts (*Giardia*, *Entamoeba coli* and *E. histolytica*) and 90–99% helminth eggs. In the case of dual-media two-layer filtration, bacterial reduction can increase from about one logarithmic unit to two to three logarithmic units.

7.8. Mobile material filters

DynaSand is a continuous sand filter developed by the Axel Johnson Institute in Sweden in the late 1970s. It is used to clean and produce drinking water, from industrial process water, with wastewater tertiary filtration. The DynaSand filter is based on the counterflow operation of the flow to be processed.

Raw water enters the upper part of the filter and is then fed to the bottom of the structure. Solids are retained in the medium as the raw water flows upwards through the filter bed. An air pump continuously draws water containing the accumulated solids from the bottom of the sand washer, where the retained solids are separated from the sand.

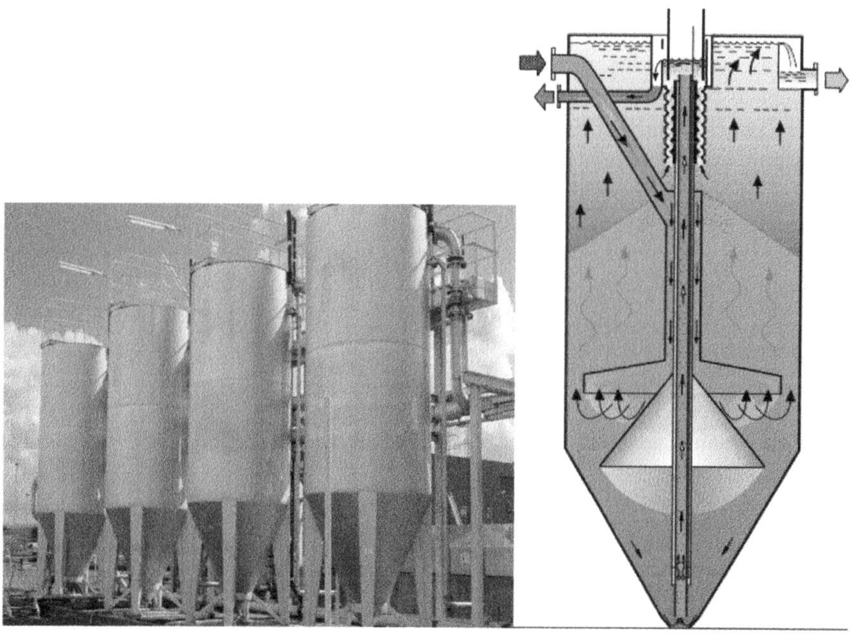

Figure 7.24. *DynaSand™ (Nordic Water). For a color version of this figure, see www.iste.co.uk/gaid/reuse1.zip*

7.9. Rotary drum filters

Tertiary treatment of wastewater is based on several methods of depollution. Among these, there is the use of rotary drum filters. These machines allow the polluted water to be filtered more finely at the WWTP level. In the same way, bacteriological treatment with UV radiation can be applied to water to remove pollutants.

The element in a disk filter is a rotating vertical disk; one or more such disks are arranged on a shaft, with roughly half of each disk submerged in a slurry tank. The device allows continuous filtration. The disks are made up of several individual sectors, each of which is an independent filter. The interior space of a sector is connected to the filtrate discharge by a rotary valve, so that a cake can form on the outside when the sector is submerged in the tank.

Also called disk filter, *the rotary drum filter* is a mechanical filtration device. Often made of stainless steel, it is composed of a cylindrical drum with a mesh size from 26 to 100 μm depending on the application. This tool is useful for optimizing the separation of liquids and solids of very small size. It is used to filter urban effluents and purify wastewater properly. Rotary drum filters can be used in WWTPs as well as for the treatment of drinking water or fish farming. The wastewater to be treated is guided through the drum and drains by gravity through the filter panels attached to the periphery of the sieve. SS are eliminated by backwashing and accumulate progressively inside the rotary drum filters through a waste collection and disposal launder.

When the difference in level between upstream and downstream is too significant, the panel turns. This starts the cleaning cycle to filter SS located on the filter cloth. The water from the cloth cleaning is then recovered and discharged independently of the filtered water.

The Hydrotech filter (Veolia) can be used in tertiary treatment: in municipal wastewater plants to treat industrial effluents (food, paper, etc.) or even in aquaculture.

This is a rotary disk filter, as each disk is covered with a filter cloth whose pore size can vary between 5 and 1,000 μm.

Water flows through the filter from the center to the ends by gravity. The purified water passes through each filter disk.

The water level gradually rises when the disks become clogged up to a predetermined threshold. The percentage of immersion is typically 55–65%. Once this level is reached, the disks begin rotating to initiate a wash cycle. Filtered water is then sent at high pressure through nozzles located transversely to each disk. Nozzles can be moved to improve filter washing. Generally, the wash requires 1–2% of the incoming flow.

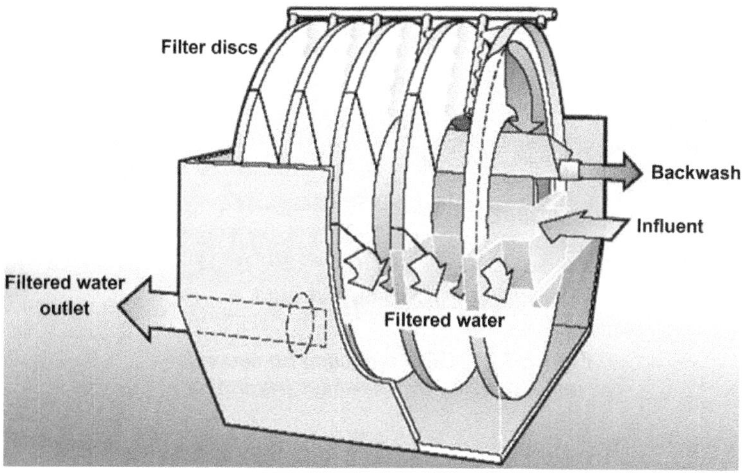

Figure 7.25. *Hydrotech filter (Veolia): disk filter. For a color version of this figure, see www.iste.co.uk/gaid/reuse1.zip*

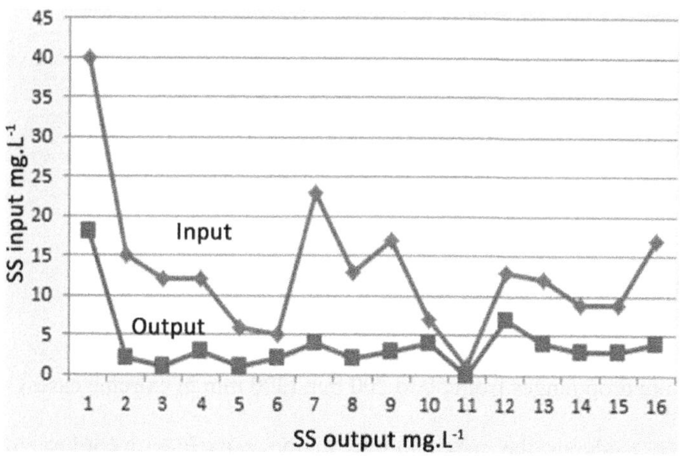

Figure 7.26. *SS reduction on secondary effluent using the disk filter (Hydrotech)*

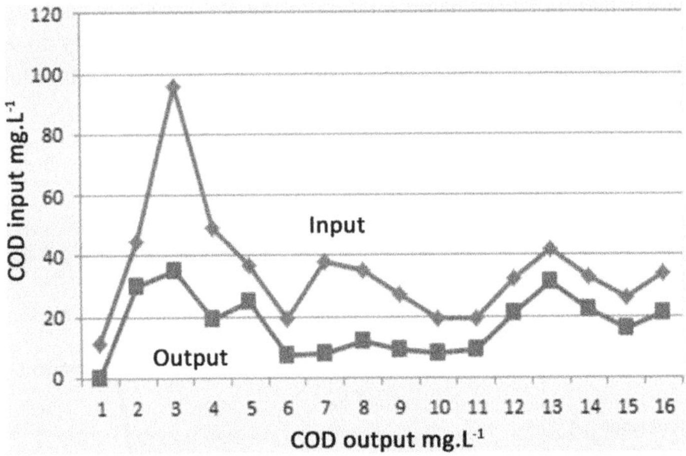

Figure 7.27. *COD reduction on secondary effluent using the disk filter (Hydrotech)*

The disk filter performance shows a reduction of 40%–91% depending on the SS input concentration.

7.9.1. *Hydrotech filter characteristics*

Depending on the model:

– the specific filtration area can vary from 8.4 to 108 m^2;

– the disk diameter can be 1.7, 2.1 and 3.1 m;

– the filter length can vary from 2.3 to 4.2 m;

– the filter can contain up to 12 disks;

– the filter material is polyester;

– pore size ranges from 5 to 1,000 μm;

– pressure drop ranges from 50 to 200 mm (300 mm in extreme cases).

Table 7.18 shows the maximum hydraulic capacities according to the disk diameter and filter material pore size for water with SS concentration less than 10 $mg \cdot L^{-1}$.

Diameter (m)	Pore size (μm)	Maximum hydraulic capacity (L·s^{-1})
1.7	10	313.2
	15–90	720
2.1	10	61.2–756
	15	180–2,016
	25*	237.6–2,880
	90*	288–2,880
3.1	10	378–1,508
	25*	1,426–5,400
	90*	1,620–5,400

*There are several pore sizes between 25 and 90 μm.

Table 7.18. *Design of the Hydrotech-type mechanic filters*

7.9.1.1. *Elimination of helminth eggs*

It has been shown that microsieves (Hydrotech) with a 10 μm cut-off mesh eliminate up to 4 log of helminth eggs.

Between 1,800 and 20,000 helminth eggs (*Trichuris suis* species) per liter were injected into treated wastewater and then fed to microsieves operating at a filtration rate of 24 m·h^{-1}. The results clearly showed that the water filtered by the microsieves did not contain helminth eggs. Between 3.5 and 4.3 reduction log were observed depending on the input concentration. Filtered water was consistently free of parasites, and the guarantee limit (<1egg/10 L) was always respected.

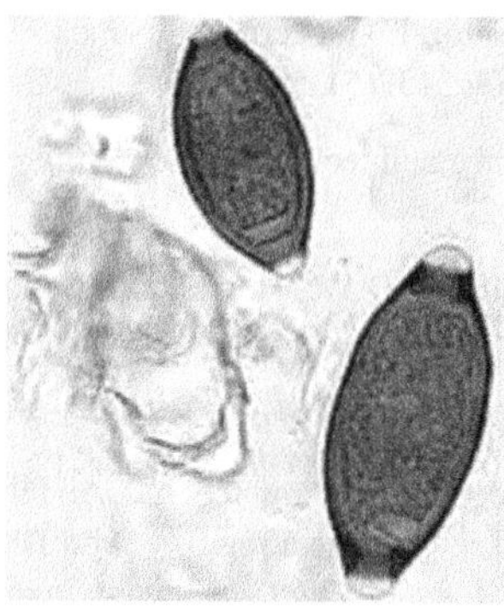

Figure 7.28. Trichuris suis *Helminth eggs. For a color version of this figure, see www.iste.co.uk/gaid/reuse1.zip*

The Hydrotech microsieve acts as an effective and reliable barrier against helminth eggs. Its small footprint, low energy consumption, and simple operation and maintenance requirements make it a competitive and attractive technology for water recovery and reuse.

7.10. References

Adin, A., Baumann, E.R., Cleasby, J.L. (1979). The application of filtration theory to pilot-plant design. *J. Amer. Water Works Assoc.*, 71(1), 17–27.

Amirtharajah, A. (1977). Optimum backwashing of sand filters. *J. Environ. Eng. Div. ASCE*, 104(EE5), 917–938.

Amirtharajah, A. and Cleasby, J.L. (1972). Predicting expansion of filters during backwash. *J. Amer. Water Works Assoc.*, 64(1), 52–59.

Armon, R., Mandelbaum, R., Manor, Y., Campos, C., Gillerman, L., Salgot, M., Gerba, C., Klein, I. (2001). Secondary wastewater disposal for crop irrigation with minimal risks. *Water Sci. Technol.*, 43(10), 139–146.

Asano, T. and Cotruvo, J.A. (2004). Groundwater recharge with reclaimed municipal wastewater: Health and regulatory considerations. *Water Res.*, 38, 1941–1951.

Camp, T.R., Graber, S.D., Conklin, G.F. (1971). Backwashing of granular water filters. *J. Environ. Eng. Div. Proc. ASCE*, 97(EE6), 903–919.

Carman, P.C. (1937). Fluid flow through granular beds. *Chemical Engineering Research & Design*, 75, S32–S48.

Carman, P.C. (1947). Slip flow in granular beds. *Nature*, 160(4061), 301–302.

Cho, A.G., Dodds, J., Santamarina, J.C. (2006). Particle shape effects on packing density, stiffness and strength – Natural and crushed sands. *Journal of Geotechnical and Geoenvironmental Engineering*, 132(5), 591–602.

Cleasby, J.L. (1993). Water science and technology. *Declining Rate Filtration, Water Science and Technology*, 277(8), 11–22.

Crittenden, J.C., Trussell, R.R., Hand, D.W., Howe, K.J., Tchobanoglous, G. (2012). *MWH's Water Treatment: Principles and Design*. John Wiley & Sons, New York.

Culp, R.L. (1977). Direct filtration. *J. Amer. Water Works Assoc.*, 69, 375–391.

Dawoud, M.A., Sallam, O.M., And Mahmoud, A., Abdelfattah, M.A. (2010). Treated wastewater management and reuse in arid regions: Abu Dhabi case study water resources department. In *Environment Agency, 10th Gulf Water Conference*, Damas.

Ergun, S. (1952a). Fluid flow through packed columns. *Chemical Engineering Progress*, 48(2), 78–89.

Ergun, S. (1952b). Determination of geometric surface area of crushed porous solids. *Analytical Chemistry*, 24, 369–388.

Ergun, S. and Orning, A.A. (1949). Fluid flow through randomly packed columns and fluidized beds. *Industrial and Engineering Chemistry*, 41(6), 1179–1184.

Exall, K. (2004). A review of water reuse and recycling, with reference to Canadian practice and potential. *Water Qual. Res. J. Canada*, 39(1), 13–28.

Exall, K., Marsalek, J., Schaefer, K. (2004). A review of water reuse and recycling, with reference to Canadian practice and potential 1. Incentives and implementation. *Water Qual. Res. J. Canada*, 39(1), 1 12.

Fasciolo, G., Meca, M., Gabriel, E., Morabito, J. (2002). Effects on crops of irrigation with treated municipal wastewaters. *Water Sci. Technol.*, 45(1), 133–138.

Fitzpatrick, C.S.B. (1990). Detachment of deposits by fluid shear during filter back-washing. *Water Supply*, 8, 177–183.

Gaid, K. (2002). Wastewater reuse for the industry. In *Conf. Environment – Industry*, Paris.

Gaid, K. and Bonne, P. (2003). Les technologies membranaires dans la réutilisation des eaux usées. *L'Eau, L'Industrie, Les Nuisances*, 125, 56–68.

Gaid, K. and Lanning, J. (2000). Wastewater reuse: The case of Durban. In *French-India Conf. on Environment*, Mumbai.

Gaid, K. and Le Poder, N. (1998). La réutilisation des eaux usées pour l'agriculture : critères de qualité et moyens techniques. In *IXe Congrès U.A.D.E.*, Casablanca.

Gaid, K. and Nugier, L. (2006). Traitement tertiaire des eaux usées en France. Le cas de l'usine de l'Hermitage (La Réunion). *L'Eau, L'Industrie, Les Nuisances*, 322, 322–342.

Gaid, K. and Zeghal, S. (1998). Wastewater technologies reuse. In *Water & Environment Conference*, Paris.

Graham, N.J.D. (1988a). Filter pore flocculation as a mechanism in rapid filtration. *Water Research*, 22(10), 1229–1238.

Hamoda, M.F., Al-Ghusain, I., Al-Mutairi, N.Z. (2004). Sand filtration of wastewater for tertiary treatment and water reuse. *Desalination*, 164(3), 203–211.

Hutchinson, W.R. (1976). High-rate direct filtration. *J. Amer. Water Works Assoc.*, 65(9), 292–298.

Kawamura, S. (1975). Design and operation of high-rate filters. *J. Amer. Water Works Assoc.*, 67(10), 535–549.

Koivunen, J., Siitonen, A., Heinonen-Tanski, H. (2003). Elimination of enteric bacteria in biological/chemical wastewater treatment and tertiary filtration units. *Water Res.*, 37, 690–698.

Kozeny, J. (1927). Uber Kapillare Leitung des Wasser in Boden. *Wien Akad. Wiss.*, 136, 271–306.

Lawler, D.F. and Nason, J.A. (2006). Granular media filtration: Old process, new thoughts. *Water Science and Technology*, 53(7), 1–7.

Leva, M. (1959). *Fluidization*. McGraw-Hill, New York.

Liao, C.W., Yu, J.H., Tang, Y.S. (2010). On-line full scan inspection of particle size and shape using digital image processing. *Particuology, Chinese Society of Particuology*, 8(3), 286–292.

McEwen, B. (1998). Indirect potable reuse of reclaimed water. In *Wastewater Reclamation and Reuse*, Asano, T. (ed.). CRC Press, New York.

Mitchell, C., Chong, J., Turner, A., Retamal, M., Carrard, N., Murta, J., Mukheibir, P., Moy, C. (2013). Darling Quarter Case Study. Building Industry Capability to Make Recycled Water Investment Decisions. Institute for Sustainable Futures, University of Technology, Sydney, for the Australian Water Recycling Centre of Excellence.

Mujeriego, R., Cazurra, T., Gullon, M. (2008). Wastewater reuse in El Prat de Llobregat, Barcelona. *Water and Sciences Technology*, 57(4), 567–574.

Ncube, P., Pidou, M., Stephenson, T., Jefferson, B., Jarvis, P. (2016). The effect of high hydraulic loading rate on the removal efficiency of a quadruple media filter for tertiary wastewater treatment. *Water Research*, 107, 102–112.

Ni, C.H., Chen, J.N., Tsai, Y.C., Chen, T.K., Chen, W.B., Chen, C.H. (2003). Recycling and reuse of wastewater from a new-developed community using sand filtration, ultrafiltration, and ozonation. *J. Environ. Sci. Health A: Tox. Hazard. Subst. Environ. Eng.*, 38(10), 2339–2348.

Ogoshi, M., Suzuki, Y., Asano, T. (2001). Water reuse in Japan. *Water Sci. Technol.*, 3(10), 17–23.

Okun, D. (1997). Distributing reclaimed water through dual systems. *J. Amer. Water Works Assoc.*, 89(11), 52–64.

Olivieri, A.W., Eisenberg, D.M., Cooper, R.C., Tchobanoglous, G., Gagliardo, P. (1996). Recycled water – A source of potable water: City of San Diego health effects study. *Water Sci. Technol.*, 33(10/11), 285–296.

Perez, S., Köck-Schulmeyer, M., Ginebreda, A. (2010). Wastewater reuse in the Mediterranean area of Catalonia, Spain. Case study of reuse tertiary effluent from a waste-water treatment plant at El Prat de Llobregat (Barcelona). In *Waste Water Treatment and Reuse in the Mediterranean Region*, Barceló, D. and Petrovic, M. (eds). Springer, Berlin.

Pollice, A., Lopez, A., Laera, G., Rubino, P., Lonigro, A. (2004). Tertiary filtered municipal wastewater as alternative water source in agriculture: A field investigation in Southern Italy. *Science of the Total Environment*, 24, 201–210.

Quinzaños, S., Strube, R., Dahl, C., Mujeriego, R. (2008). Helminth eggs removal by micro screening for water reclamation and reuse. *Water Science & Technology*, 57(5), 715–720.

Rodgers, S.E. and Lauer, W.C. (1992). Denver's demonstration of potable water reuse: Water quality and health effects testing. *Water Sci. Technol.*, 26(7/8), 1555–1564.

Sanz, J., Strube, R., Quinzaños, S., Dahl, C., Montoliu, I., Gracenea, M., Mujeriego, R. (2009). Helminth eggs removal in water reclamation on disk-filters. In *International Conf. on Disinfection of Wastewater*, Dubai, 1–14.

Sneed, E.D. and Folk, R.L. (1958). Pebbles in the Lower Colorado River, Texas: A study in particle morphogenesis. *Journal of Geology*, 66(2), 114–150.

Stefanakis, A. (2015). Modern water reuse technologists: Tertiary membrane and activated carbon filtration. In *Urban Water Reuse Handbook*, Eslamian, S. (ed.). CRC Press, New York.

Tchobanoglous, G., Burton, F.L., Stensel, H.D. (2002). *Wastewater Engineering Treatment and Reuse*. McGraw-Hill, New York.

Wen, C.Y. and Yu, Y.H. (1966). Mechanisms of fluidization. *Chemical Engineering Progress*, 62(4), 75–85.

Yao, K.M., Habibian, M.T., O'Melia, C.R. (1971). Water and wastewater filtration: Concepts and applications. *Environ. Sci. Technol.*, 5, 1105–1112.

Zhou, H. and Smith, D.W. (2002). Advanced technologies in water and wastewater treatment. *Journal of Environmental Engineering and Science*, 1, 247–264.

8

Disinfection

One of the most critical steps in any reuse program is to protect public health, especially the health of workers and consumers. For this purpose, it is very important to neutralize or eliminate any infectious agents or pathogens that may be present in wastewater.

Treated wastewater reuse requires a water quality sufficient for the intended use. In wastewater treatment plants, pathogens (bacteria, viruses, parasites and helminth eggs) are present at concentrations well above the WHO guideline values and quality parameters listed in national standards of countries who reuse treated wastewater. To protect the health of people in direct contact with treated wastewater, such as farmers and irrigation technicians, as well as various users (consumers, industrialists, hydraulic engineers, etc.), disinfection is necessary.

The final stage of wastewater tertiary treatment before reuse is to eliminate all pathogens through the disinfection process. The purpose of disinfection is therefore to distribute water with a microbiological quality in line with the various uses. This quality may be of drinking water type for water supplies and food industries but may be less restrictive for other uses. Several disinfection agents can be used, such as chlorine, ozone and ultraviolet (UV).

8.1. Microorganisms present in treated wastewater

The main water-borne diseases are infections caused by the following agents: bacteria, viruses and parasites. The WHO (World Health Organization) provides an exhaustive list. Apart from enterobacteria (salmonella, shigella, *E. coli*, *Campylobacter*, *Pseudomonas aeruginosa* and *Vibrio cholerae*), it contains viruses (adenovirus, enterovirus, rotavirus and hepatitis virus) and, above all, parasites (*Cryptosporidium* and *Giardia*), whose importance is increasingly being taken into account. Pathogenic organisms found in wastewater may be excreted by human beings and animals who are infected with disease or who are carriers of a particular infectious disease. The pathogenic organisms found in wastewater can be classified into four categories: bacteria, protozoa, helminths and viruses.

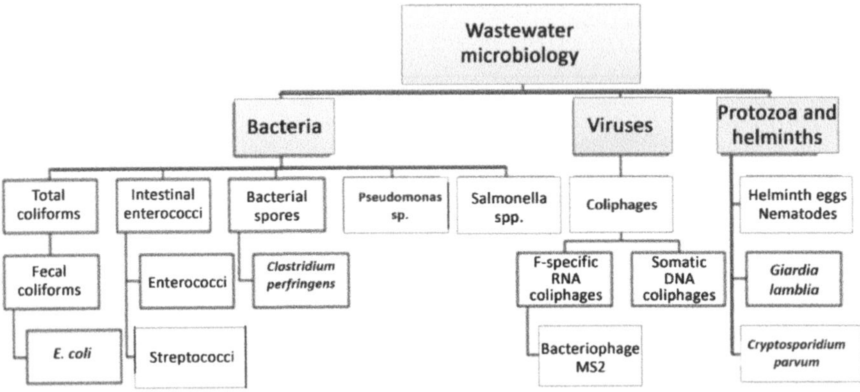

Figure 8.1. *Wastewater microbiology*

8.1.1. *Bacteria*

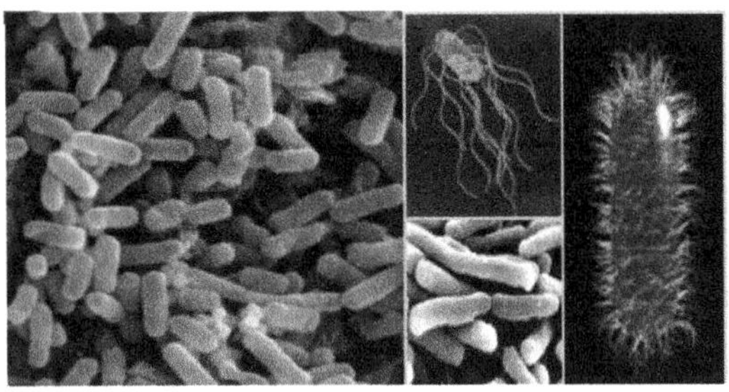

Figure 8.2. *Various species of bacteria. For a color version of this figure, see www.iste.co.uk/gaid/reuse1.zip*

Bacteria are intermediate in size between that of protozoa and viruses and are 1–10 μm long. They are single-celled organisms, but unlike protozoa, their DNA is not enclosed in a nucleus. Bacteria can be of various forms (sticks: bacilli; spheres: cocci), they are present in all environments and have a multitude of different lifestyles. In addition to the natural and harmless bacteria, water contains pathogenic species, most often of fecal origin.

Because pathogenic bacteria are present in the feces of infected individuals, domestic wastewater contains a wide variety and concentration range of non-pathogenic and pathogenic bacteria.

The most common species are *Salmonella* and *Shigella*, as well as *Escherichia coli*, *Vibrio cholerae*, *Yersinia enterocolitica* and *Campylobacter fetus*.

Bacterial outbreaks are the majority of water-borne epidemics:

– *Legionella pneumophila* (pneumonia).

– *Shigella* (gastrointestinal disorders).

– *Salmonella* (typhoid and paratyphoid fevers, salmonellosis) are responsible for a multitude of syndromes. Frequently responsible for food poisoning, they are also present in water.

– *Pathogenic staphylococci*: these germs are the cause of a very wide variety of infections: skin conditions, impetigo, septicemia, endocarditis and osteomyelitis.

– *Enteroviruses*: abundantly excreted with fecal matter, they may contaminate supply water. The main enteroviruses are polioviruses (responsible for poliomyelitis), coxsackieviruses and echoviruses. Overall, they may induce a wide range of syndromes including skin rashes, fever, gastroenteritis, myocarditis, meningitis, respiratory diseases and hepatitis.

– *Coliforms*: coliform bacteria exist in fecal matter, but also grow in natural environments: soil, vegetation and natural water. Their concentration in treated wastewater is between 10^5 and 10^7 u/100 mL.

– *Thermotolerant coliforms*: these coliforms are able to grow at 44°C, whereas no growth is observed at this temperature for non-fecal strains. The main coliform bacteria specifically of fecal origin is *Escherichia coli*. These bacteria still appear in large quantities in animal and human excreta.

– *Fecal streptococci*: they are normal hosts of human and warm-blooded animal intestines. Fecal streptococci belong to a group of streptococci that are not all fecal. However, their research in conjunction with fecal coliforms is a good indicator of fecal contamination. Some studies have even shown that streptococci are a better indicator than fecal coliforms for certain water-borne infectious diseases.

– *Spores of sulfate-reducing bacteria*: these are resistant ("sporulated") forms of anaerobic organisms, the most common and easiest to identify being clostridia (sulfate-reducing *Clostridium*). They are normally present in fecal matter, but in smaller quantities than *E. coli*. Similarly, the absence of spores after a filtration treatment is a sign of the latter's effectiveness. Their persistence in disinfected water may be a sign of defects in physicochemical treatments, lack of effectiveness of certain disinfectants or insufficient filtration. Because of their similar behavior to parasites (formation of resistant cysts), spores are a good indicator for these microorganisms. This explains why it is not the bacteria themselves that are being sought, but their spores.

8.1.2. *Viruses*

These microorganisms are more resistant than bacteria. They may exist in a vegetative state, or be active on living cells:

– hepatitis A virus;

– polio virus;

– conjunctivitis virus.

8.1.2.1. *Fecal bacteriophages*

They are specific viruses of fecal bacteria corresponding to them. Bacteriophages are viruses that infect bacteria as the host. They are found in water contaminated by fecal matter. They are therefore a good indicator of these contaminations.

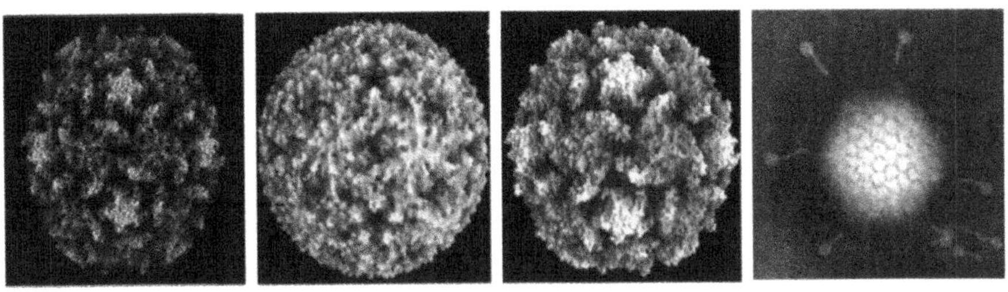

Figure 8.3. *A few surface-water viruses. For a color version of this figure, see www.iste.co.uk/gaid/reuse1.zip*

Their presence may indicate past or intermittent contamination. Thanks to their resistance to chlorinated treatments, they can be detected in a network and are the only witnesses of previous contamination. They are not considered pathogenic microorganisms.

With a size from 10 to 300 nm (1 nm = 10^{-6} mm), invisible under an optical microscope, viruses do not have a cell structure but are in some way reservoirs of genetic material. Viruses cannot live independently and can only develop as parasites of living cells (bacteria, protozoa, higher organisms, etc.). They use the biochemical machinery of their host to their advantage. Viruses are composed of a nucleic acid core (either DNA or RNA) surrounded by an outer shell of protein called a capsid. They are obligate intracellular parasites that multiply only within a host cell, where they redirect the cell's biochemical system to reproduce themselves.

The presence of viruses in treated wastewater is very frequent; some have a shell that makes them very resistant in aquatic environments. This is particularly the case of human digestive tract viruses responsible for polio and hepatitis A, which are very common in wastewater and treated wastewater.

8.1.3. *Parasites*

The main parasites are protozoa and helminths (worms). There are many helminths that can contaminate wastewater. Most helminths fall into three major types: Nematoda (roundworms), Platyhelminthes (Flatworm) and Annelida (segmented worms). Most human infections are associated with nematodes and flatworms.

One of the best known is responsible for bilharzia, a serious urinary, digestive or arteriovenous disease, common in developing countries. Helminth eggs, which range in size from about 10 µm to more than 100 µm, can be removed by several wastewater processes such as sedimentation and stabilization ponds. These worms and their eggs are not destroyed by disinfection, but they are well eliminated by filtration and very well by membranes.

Protozoa are mobile single-celled organisms whose DNA is contained in a nucleus surrounded by a membrane. Their cysts are among the most resistant forms to wastewater treatments, including chemical disinfection. The cysts are *Giardia lamblia*, and the oocysts of *Cryptosporidium* are the most resistant forms. *Cryptosporidium* is 4–6 µm in size and *Giardia* is 8–10 µm. These organisms are a particular concern because they are found in almost all wastewater and because conventional disinfection techniques using chlorine have not proved to be effective in their inactivation.

Protozoa feed on suspended organic particles or bacteria and most often reproduce asexually (the division of one "mother" cell gives birth to two "daughter" cells). There are four main classes, which are differentiated by the nature of the locomotor system and by the characteristics of their lifecycle. These are:

– flagellates, which possess a flagellum (*Giardia lamblia*, etc.);

– rhizopods (*Entamoeba histolytica* amoeba, etc.);

– ciliates;

– sporozoan; their lifecycle goes through a spore stage (*Cryptosporidium*, etc.).

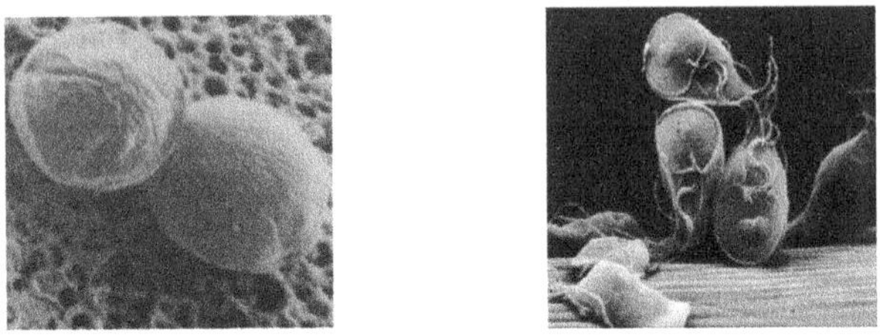

a) b)

Figure 8.4. *Cryptosporidium (a) and Giardia lamblia (b)*

8.2. General rules of chemical disinfection

Wastewater disinfection involves two important steps, corresponding to two different effects of a given disinfectant:

– germicidal effect (bactericidal and virucidal): the ability of disinfectants to destroy germs in a treatment step of the process;

– residual effect: it is an effect of the disinfectant that remains in the produced wastewater and ensures the microbiological quality of the water transported. It is both a bacteriostatic effect against bacterial regrowth and a germicidal effect against weak and punctual pollution that may occur in the pipes or in the reservoirs of produced treated wastewater.

8.2.1. *Mode of action of chemical disinfectants in treated wastewater*

The various disinfection techniques are:

– chlorine disinfection:

- sodium hypochlorite,
- chlorine gas;

– disinfection with chlorine dioxide;

– chloramine disinfection;

– UV disinfection typically associated with chlorine post-disinfection;

– ozone disinfection combined with chlorine post-disinfection.

The disinfectants commonly used have a highly effective germicidal power, essentially based on chemical mechanisms of oxidoreduction. Thus, the effectiveness of a disinfectant is directly related to its oxidizing power, which makes it possible to classify reagents in descending order of efficiency according to their redox potential (Table 8.1).

Radical	**$\cdot OH$**	**2.80 V**
Ozone in acidic medium	O_3/O_2 (H^+)	2.07 V
Peracetic acid	CH_3COOOH	1.96 V
Hypochlorous acid	HClO/ClO	1.49 V
Chlorine	Cl_2/Cl	1.36 V
Ozone in alkaline medium	O_3/O_2 (OH)	1.24 V
Chlorine dioxide	ClO_2/ClO_2	1.27 V
Hypochlorite	ClO/Cl	0.90 V

Table 8.1. *Redox potential of various oxidants*

It was observed that wastewater treatment plants with similar water qualities in terms of BOD, COD and nitrogen responded differently to disinfection by chlorination. This may be explained by the presence of interfering organic compounds in reaction mechanisms. These compounds are distinguished by their functional groups and chemical structures. Saturated compounds and carbohydrates have little or no influence on chlorine demand and therefore do not interfere with the disinfection process.

Organic compounds with unsaturated bonds, on the other hand, have an immediate demand for chlorine, depending on their functional groups. Aromatic ring compounds containing hydroxyl functions and those with sulfide functions react directly with chlorine to form compounds that have little or no bactericidal potential. These cases require higher doses or longer contact times.

The effectiveness of various disinfectants will be influenced by the type, nature and condition of the microorganisms. For example, growing viable bacterial cells are often killed more easily than older cells that have developed a viscous coating. Bacterial spores are extremely resistant, and many chemical disinfectants normally used will have little or no effect. Similarly, many viruses and protozoa of concern react differently to each chemical disinfectant. In some cases, other disinfectants such as UV rays should be used.

The sites of action of the disinfectant depend on the nature of the target microorganism and its chemical structure. The five main mechanisms proposed to explain the action of disinfectants are as follows: damage to the cell wall, alteration of cellular permeability, alteration of the colloidal nature of protoplasm, alteration of DNA or RNA organisms and inhibition of enzyme activity.

In the case of viruses, the oxidant alters complex macromolecules of nucleic acids (DNA and RNA), thus preventing the microorganism from reproducing and reducing its infectivity. In bacteria, the oxidant penetrates the cytoplasmic membrane and oxidizes the sulfur bonds of enzymes, irreversibly altering them, leading to the death of the microorganism. Higher doses of oxidant also lead to degradation of the bacteria's nucleic acids.

Viruses are naturally more resistant than bacteria, so virucidal treatment conditions will be more severe than bactericidal treatment conditions, regardless of the disinfectant used.

Some microorganisms can adopt life forms that are highly resistant to chemical attack. Thus, bacteria that have survived boundary conditions (nutrient deprivation, repeated oxidation, or temperature) show an increased resistance to the action of oxidants, compared to the same strains subjected to optimal growth conditions.

Sporulation mechanisms (secretion by the micro-organism of a protective polysaccharide "sheath") allow some germs to withstand boundary conditions by adopting slowed-down life forms.

UV disinfection generally involves the use of the absorption spectrum of UV radiation by microorganisms, which is between 200 and 300 nm. The maximum absorption, corresponding to a maximum biocide action, is obtained for a wavelength of 253.7 nm.

UV radiation acts on the DNA and RNA of microorganisms by blocking their replication, which stops cell division. Prolonged irradiation causes the cell nucleus to dilate, leading to the cell's rupture.

8.2.2. *Reduction principle of the number of germs – "Log" units*

The concept of log unit was first used by US authors. It makes it possible to turn the reduction in water turbidity following a clarification step into the percentage of elimination (then called reduction) of suspended particles. Particle reduction, expressed in log units, is calculated according to the following formula:

$$\text{Number of reduction} = \log \frac{\text{Initial concentration}}{\text{Final concentration}}$$

$$\text{Reduction \%} = \frac{\text{Initial concentration} - \text{Final concentration}}{\text{Initial concentration}}$$

For example, if the particle concentration has been reduced by a factor of 100 following treatment:

– the reduction percentage is 99%;

– the reduction in log units is equal to 2.

Similarly, 3 log reduction correspond to 99.9%, 4 log correspond to 99.99%, etc. This particle reduction measure also applies to microorganisms suspended in water. It therefore makes it possible to measure the physical retention following a type of treatment and, consequently, to compare the respective efficiencies of several systems. Initially applied to physical retention processes, this unit is now used by extension to measure the effectiveness of chemical disinfection treatments (we then speak of level of inactivation).

8.2.2.1. *Ct notion*

Chemical disinfection of wastewater intended for consumption relies on the use of oxidizing properties of various chemical reagents. The most widely used currently are chlorinated compounds (chlorine gas, sodium and calcium hypochlorite), chlorine dioxide and ozone.

Microorganisms have different sensitivities to disinfectants, which themselves have distinct oxidative powers. For a given micro-organism, the effectiveness of disinfection is affected in particular by the disinfectant concentration, contact time, pH, temperature, as well as water disinfection demand. The product of the disinfectant concentration (C in $mg \cdot L^{-1}$, measured at the outlet of the contact tank) and time (t in minutes) is called Ct (in $mg \cdot L^{-1} \cdot min^{-1}$).

The higher the Ct value, the more effective the disinfection. The concentration or contact time, or both, can be modified to obtain the desired and most effective Ct value.

Ct values are useful for comparing the effectiveness of disinfectants. The lower the Ct value to achieve a given level of microorganism inactivation, the more effective the disinfectant. The Ct value is also used to classify the relative sensitivity of different microorganisms. The higher the Ct value required to achieve a given level of inactivation, the more resistant the microorganism.

For each compound, the germicidal power depends on both the residual concentration in disinfectant (C) and the contact time (t), hence the Ct notion.

The effectiveness of a disinfection can be equated by Chick's law:

$$\frac{Nt}{No} = exp\ (-kt)$$

and:

$$k = \lambda C^{\eta}$$

with:

– N_0: number of microorganisms initially present;

– Nt: number of microorganisms surviving at time t in the presence of concentration C;

– C: disinfectant concentration (assumed to be constant over time);

– k: rate constant, depends on the microorganism, disinfectant and operating conditions (temperature, pH, ionic strength of the medium);

– t: contact time with the disinfectant;

– λ: lethality coefficient of the microorganism in the presence of the disinfectant;

– n: constant called "dilution coefficient", close to 1.

This law reflects first-order kinetics in relation to time, and of n order in relation to the oxidant concentration for a given microorganism, under given operating conditions and for a constant concentration of oxidant. Doubling the contact time should therefore make it possible to reduce an additional logarithmic scale of germs. This law highlights the importance of the Ct product, the product of the C disinfectant concentration by the time of contact t (Chick 1908; Watson 1908).

Thus, disinfection is considered effective when a C residual oxidant is maintained for a time t in the water to be treated. For a given disinfectant, such as ozone or chlorine, the Ct required for inactivation of n log germs depends on the target microorganism as well as temperature and pH. For the same microorganism, the Ct depends on the oxidant used and the operating conditions (pH and temperature).

In practice, C is similar to the residual disinfectant concentration at the outlet of the disinfection tank. The precise determination of t can only be done with a tracer. It is often recommended to use the t_{10} value in Ct calculations.

This is the time at which 10% of the flow exits the reactor (i.e. minimum transit time of 90% of the water flow). In practice, t represents 25–50% of the average dwell time, obtained by dividing the volume of the tank by the through-flow:

$$t = \frac{\text{Disinfection tank volume}}{\text{Flow}}$$

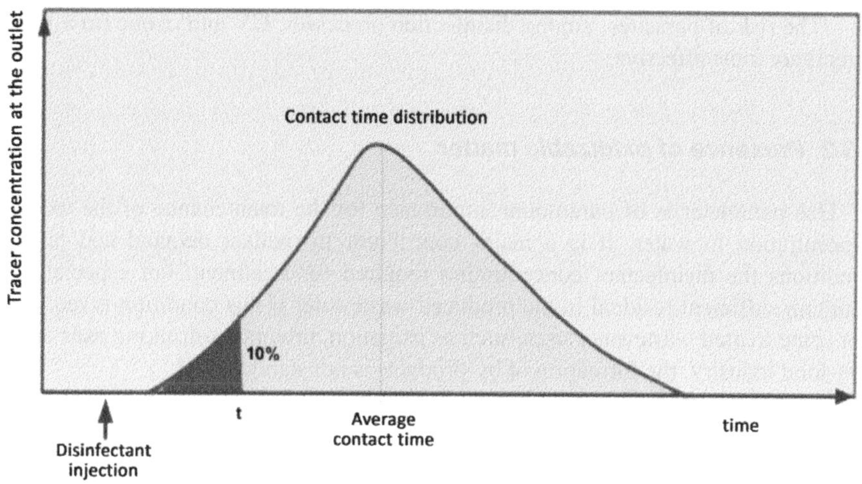

Figure 8.5. *Dwell time distribution in a disinfection tank*

Typically, the chlorination contact time is based on peak flow. The dose typically ranges from 1 to 3 mg·L^{-1}. There are several other factors that ensure optimal conditions for disinfection, and they include temperature, suspended solids (SS), turbidity, organic matter, pH and oxidizable compound concentration (NH_4^+, NO_2^-, Fe, Mn, color, etc.).

8.3. Factors influencing the effectiveness of chemical disinfection

The following factors influence the performance of chemical disinfection.

8.3.1. *Turbidity*

Turbidity related to the SS in the treated wastewater before disinfection should be as low as possible for the following reasons:

– Colloidal turbidity, as well as SS, are actually attachment supports for germs ("shelters"), which reduces the effectiveness of disinfectants. In the case of UV disinfection, turbidity should be as low as possible, since it determines the UV absorption coefficient by the wastewater to be disinfected and therefore the effectiveness of the radiation on germs. SS protects microorganisms from the action of disinfectants.

– The risk of parasites: among disinfection processes, UV and ozone (to a lesser extent) are most effective.

8.3.2. *Presence of oxidizable matter*

This parameter is of paramount importance for the maintenance of the oxidant concentration in water. It is a major contributor to oxidant demand and largely conditions the disinfectant concentration required for treatment, but especially to maintain sufficient residual in the produced wastewater if this condition is required. For some treated wastewater uses, such as irrigation, urban non-drinking uses or the non-food industry, the formation of by-products is not standardized.

8.3.3. *Injection mode and point*

The disinfectant must be mixed closely and quickly with the wastewater to be treated for optimal efficiency. To ensure disinfection, the reagent will be injected upstream of a tank leaving sufficient contact time.

For specific cases where the disinfection treatment includes two reagents operating in a complementary manner, such as ozone followed by chlorination, chlorine should not be injected into wastewater containing a residual ozone, as there is competition between these two oxidants, leading to an overconsumption of reagent. For the same reasons, chlorine dioxide should not be injected into wastewater containing residual ozone. The reaction between ozone and chlorine dioxide also leads to the formation of chlorates, which are undesirable in sanitary terms.

The solution is to inject chlorine or chlorine dioxide as far downstream as possible from the ozonation towers or, if necessary, neutralize the residual ozone.

8.3.4. *Design of the contact tank*

From a hydraulic point of view, the contact tank must be designed to limit the presence of short circuits through which some water could escape the minimum required contact time. It is also necessary to restrict circulation which limits the useful volume of the tank. The interior design must therefore be optimized with weirs or baffles when the amount of treated water to store is significant.

8.4. Qualities of a good disinfectant

An ideal disinfectant should have the following qualities:

– reduce pathogenic germs, including the most resistant (cysts, spores), according to the final wastewater qualities required;

– have a residual effect;

– be stable (to avoid self-decomposition phenomena);

– not be a producer of unwanted by-products;

– inhibit biofilm growth;

– be simple and safe to implement.

In practice, each disinfection technique has advantages and disadvantages:

– *Ozone* is a powerful oxidant, and therefore an excellent germicide. However, its stability in wastewater is low, which makes any residual action impossible (ozone is considered to be stable in water for about 20 min). Ozone breaks down organic compounds into small molecules that can serve as a carbon substrate for microorganisms (CBOD), especially in distribution pipes.

– *Chlorine* is the most widely used disinfectant, because it provides both a germicidal and a residual effect, and because its conditions of application are well known and controlled. Like all oxidants used in disinfection, chlorine participates in competitive reactions with organic matter that increase the chlorine demand. By reacting with humic acids, it gives rise to halogenated compounds or "haloforms", suspected of being carcinogenic (THM).

– *Chlorine dioxide* has superior disinfectant qualities than chlorine, including on sporulated microorganisms. It does not produce sapid compounds and does not generate haloforms. The application of chlorine dioxide is quite delicate and sometimes constitutes an obstacle to its use: in fact, given its instability, dioxide cannot be transported and must therefore be prepared in situ before use. ClO_2 does not react with ammonia.

– Of all the chlorine-based disinfectants, *monochloramine* obtained by the joint use of ammonia (NH_3) and chlorine has the lowest disinfectant power. It is about 50 times less effective than chlorine and chlorine dioxide (for comparable contact times). In conventional tertiary treatment systems, there is always a more or less significant residual ammonia (NH_4^+). Thus, at the usual doses, chlorine injection systematically leads to the formation of chloramines. These are only broken down by excess chlorine. One of the most important qualities of monochloramine is its stability both in time and to heat (water 25°C and above), which gives it an interesting residual power.

Table 8.2 summarizes the qualities of the main disinfectants used in wastewater treatment.

	Ozone	Chlorine	Chlorine dioxide	Chloramines	UV
Germicidal effect	++++	++	+++	+	++
Residual effect	0	++	++	+++	0

Table 8.2. *Qualities of the main disinfectants*

8.5. The main techniques of wastewater chemical disinfection

8.5.1. *Chlorine disinfection*

The main chlorinated disinfectants used to treat wastewater are chlorine (Cl_2), sodium hypochlorite (NaOCl) and chlorine dioxide (ClO_2).

Chlorine gas and chlorinated compounds are widely used for the disinfection of water and wastewater. In solution, chlorine can present itself in different states and react with certain organic or mineral molecules.

The action of chlorine on microorganisms is on the surface by slow diffusion. When they form clusters or are attached to SS, the action of chlorine is much less effective as it does not fully spread within the formed clusters. In addition, microorganisms can release polysaccharides that create a kind of screen that may reduce the effectiveness of chlorination.

In disinfection, chlorine can be used in three forms:

– chlorine gas, Cl_2, stored under pressure in liquid phase (bottles, tanks);

– sodium hypochlorite, or NaOCl (liquid bleach);

– calcium hypochlorite (CaOCl).

The first two forms, chlorine gas and liquid bleach, are extremely common in water treatment. In all three cases, the disinfectant agent resulting from the contact of the reagent with the water to be treated is the same: it is hypochlorous acid ($HClO$) and hypochlorite (ClO^-).

The sum of both compounds is called free chlorine. Particularly in wastewater, some of the chlorine reacts with organic matter present and ammoniacal nitrogen to form chloramines or combined chlorine. The sum of free chlorine + combined chlorine is called total chlorine. The germicidal form is always $HClO$.

Generally speaking, the effectiveness of chlorine highly depends on the water pH. This environmental parameter then determines the nature of the species present. Concentrations used in practice are:

– the free chlorine content, sum of hypochlorous acid and hypochlorite concentrations;

– the active free chlorine content in the form of hypochlorous acid;

– the chlorometric degree of a hypochlorite solution, representing the concentration of a solution that releases 1 L of chlorine gas under normal conditions of temperature and pressure by addition of hydrochloric acid: 1° chlorometric is equivalent to 3.17 $g \cdot L^{-1}$ of active chlorine;

– the active chlorine content is expressed as the hypochlorite free chlorine equivalent concentration: 1 g of sodium hypochlorite equals 0.95 g of active chlorine.

<table>
<tr><th colspan="3">Forms of chlorine present in water</th><th>Composition</th></tr>
<tr><td rowspan="3">Total chlorine</td><td rowspan="2">Free chlorine</td><td>Active free chlorine</td><td>Elementary chlorine Cl_2
Hypochlorous acid ($HClO$)</td></tr>
<tr><td>Potential free chlorine</td><td>Hypochlorite ClO^-</td></tr>
<tr><td>Combined chlorine</td><td>–</td><td>Mono-, di- and trichloramines</td></tr>
</table>

Table 8.3. *Different forms of chlorine*

8.5.1.1. *Gas chlorine*

At atmospheric pressure, chlorine is a greenish-yellow gas 2.5 times denser than air, with a characteristic irritating odor. It is stored in liquid form, in bottles of 50 kg or in tanks of 1,000 kg, under a pressure of about 5 bar.

Chlorine gas is dissolved in service water, using a hydro-ejector, at an average concentration of 1 to 2 $g \cdot L^{-1}$. The solution thus obtained is chlorinated water or "chlorine water", which will be used to dose the disinfectant.

The most active form of disinfection is hypochlorous acid HOCl, called "active chlorine". Depending on the pH, hypochlorous acid is in equilibrium with the hypochlorite ion (ClO–) according to the reaction:

$$Cl_2 + H_2O \rightleftarrows \text{HClO (hypochlorous acid)} + H^+ + Cl^-$$

This reaction is accompanied by an acidification of the medium. The lower the pH, the lower the proportion of hypochlorous acid.

Active chlorine is the concentration of a potentially reactive chlorine-based compound expressed in terms of Cl_2. For example, 1 g NaOCl corresponds to 1/74.5 = 0.0134 mole of NaClO or 1 × 0.0134 × 71 = 0.95 g of active chlorine. Unlike combined chlorine, free chlorine is the result of an analysis.

For a given free chlorine content, the lower the pH and temperature (optimal at pH 5), the higher the percentage of active free chlorine formed. The pH of the wastewater to be disinfected is an essential parameter in a chlorine treatment, since the pH range corresponding to high HOCl levels is more in neutral to acid pH than in alkaline pH. Thus, at pH 8, the percentage of HOCl is only about 20% of the free chlorine measured.

Figure 8.6. *Installation of gas chlorine. For a color version of this figure, see www.iste.co.uk/gaid/reuse1.zip*

8.5.1.2. *Hypochlorite*

Two types of hypochlorite are the most used: sodium hypochlorite and commercial liquid bleach:

– sodium hypochlorite (NaOCl) is a liquid form and contains up to 12% chlorine;

– commercial liquid bleach is a sodium hypochlorite solution with a concentration of 3–5%.

Hypochlorite reacts with water to form hypochlorous acid. The reactions are as follows:

$$NaOCl + H_2O \rightleftarrows HOCl + NaOH$$

$$\text{Sodium hypochlorite + water} \rightleftarrows \text{hypochlorous acid + sodium hydroxide}$$

and:

$$Ca(OCl)_2 + 2H_2O \rightleftarrows 2HOCl + Ca(OH)_2$$

$$\text{Calcium hypochlorite + water} \rightleftarrows \text{Hypochlorous acid + lime}$$

Liquid bleach also results in the same reactions:

$$NaOCl + H_2O \rightleftarrows HClO + NaOH$$

This reaction is accompanied by a pH increase, which can lead to scaling if the dissolution water is hard.

Industrial liquid bleach "47/50" is at 47–50 chlorometric degrees, namely about 150 g of active chlorine per liter (1 chlorometric degree being equivalent to 3.17 g of active chlorine). This strength decreases in case of prolonged storage (decrease of 40% after a few months of storage or less when the outside temperature is high). Dilution in hard water can cause scaling, which can lead to degradation of the quality of the treated wastewater, pH instability and can also disturb the operation of injection pumps. Liquid bleach is corrosive and requires safety measures, especially for staff. It can be delivered in carboys, containers or tankers. It is then stored in plastic tanks inside a sealed retention casing and sheltered from light to prevent its decomposition. Whatever the reagent that produces it, hypochlorous acid is a weak acid partially dissociated in water, and is in equilibrium with the hypochlorite ion according to the reaction:

$$HClO \rightleftarrows ClO^- + H^+$$

The influence of pH on the distribution between hypochlorous acid and hypochlorite ion is shown in Table 8.4 and represented in Figure 8.7.

pH	% of total chlorine in HClO form	% of total chlorine in ClO^- form
6.0	96.5	3.5
7.0	72.5	27.5
7.2	63.5	36.5
7.3	58.6	41.4
7.4	53.3	46.7
7.5	47.0	53.0
7.6	41.1	58.9
7.7	35.5	64.5
7.8	30.0	70.0
7.9	24.5	75.5
8.0	22.0	78.0
8.2	17.0	83.0

Table 8.4. *Influence of pH on hypochlorous acid dissociation at 20°C*

Although hypochlorite has germicidal properties, the real disinfectant is hypochlorous acid. One explanation is that bacterial membranes are lipid media in which ions are not soluble (ClO–), whereas the chemical structure of hypochlorous acid allows it to penetrate the cytoplasmic membranes of microorganisms.

Hypochlorous acid is often referred to as "active free chlorine", as opposed to free chlorine in reserve in the form of hypochlorite ClO– and combined chlorine including chloramines (mono-, di- and trichloramines resulting from the action of chlorine on ammonia, and organic chloramines formed by the action of chlorine on certain organic molecules):

Total chlorine = free chlorine + combined chlorine

Combined chlorine = chloramine-bound chlorine

At a pH > 5, there is no elemental chlorine Cl_2 in water.

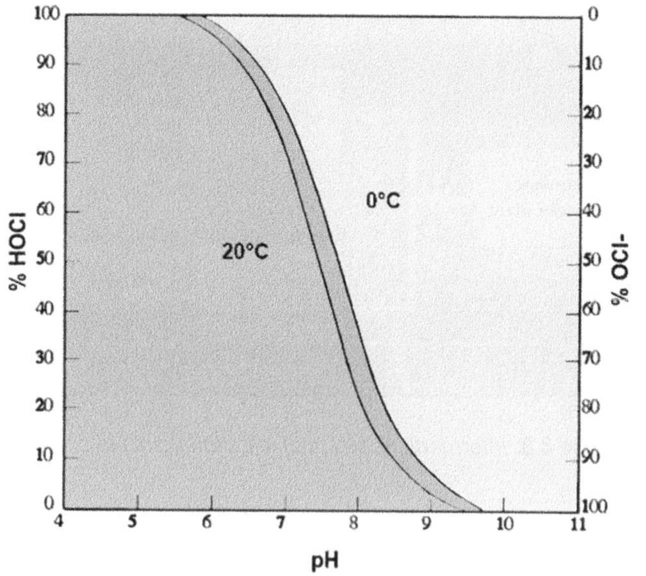

Figure 8.7. *Distribution of hypochlorous acid HClO and hypochlorite ion ClO⁻ according to pH and temperature*

8.5.1.3. *Chlorine application points*

Chlorine application points in disinfection are located downstream of tertiary filtration and upstream of contact tanks. When there is no tertiary filtration, the oxidant is injected upstream of the contact tank.

Concentrations are in the range of 5–15 mg·L^{-1} depending on the quality of the inlet water. A second injection is applied at the outlet of the contact tank, when high residual chlorine (≥1 mg·L^{-1}) is required. The main problem with the distribution of treated wastewater is the potential degradation of microbiological quality, especially in warm climates and distribution networks of significant length. For this reason and for high health risk uses, it is recommended to maintain a residual chlorine to prevent bacterial growth. However, the residual chlorine concentration should not exceed 0.5–1 mg·L^{-1} to avoid toxic effects on sensitive vegetables.

In Middle East countries adopting tertiary treatment, the trend is to introduce pre-chlorination before rapid gravity sand filtration and post-chlorination afterward. Disinfection normally involves the injection of a chlorinated solution at the top of a chlorine contact tank.

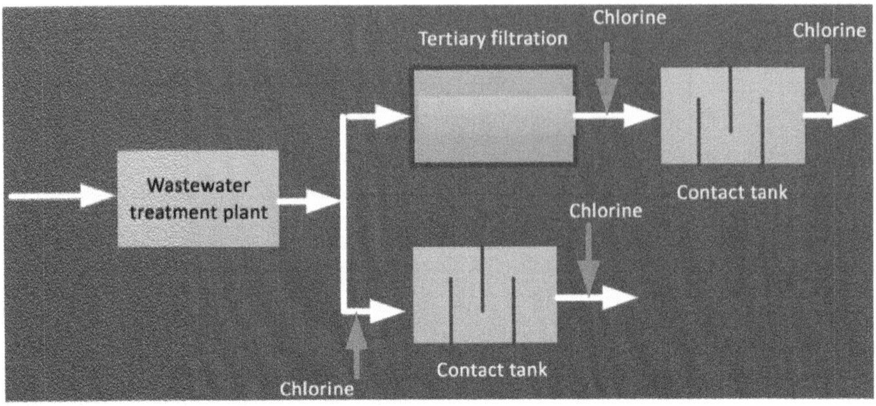

Figure 8.8. *Intermediary and final injection points of chlorine*

Water circulation in disinfection tanks directly influences the water/chlorine contact time and conditions the quality of disinfection. The contact time depends on the geometry of the tanks and their hydraulic load and has a great influence on the treatment quality. In order to obtain the optimal contact time, one approach is to simulate hydraulics using Fluent-type CFD models. This allows different situations to be considered in terms of flows, water heights and interior design. From a hydraulic point of view, it is recommended to provide for inner baffles to ensure plug flow. The aim is to maintain a reservoir in which water moves evenly from inlet to outlet, while limiting recirculation phenomena.

8.5.1.4. *Chlorine demand* – break-point

Chlorination is the process of meeting the demand for chlorine in the wastewater to be treated, that is, injecting a quantity of chlorine (or liquid bleach) sufficient to detect the appearance of free chlorine, necessary to obtain a biocidal action.

The demand for chlorine from chemical sources, which is rapid within minutes of the introduction of chlorine, slows down after a few hours.

In the absence of ammonia or nitrogenous organic matter, wastewater chlorination is very simple. Total chlorine is always equal to the available free chlorine.

In the presence of ammonia, chlorine reacts with ammonia to form a class of compounds called chloramines. Chloramines are characterized by the presence of one or more N-Cl bonds. Three types of chloramines can be formed from the reactions of chlorine and ammonia nitrogen:

$$HOCl + NH_3 \rightarrow NH_2Cl \text{ (monochloramine)} + H_2O$$

$$HOCl + NH_2Cl \rightarrow NHCl_2 \text{ (dichloramine)} + H_2O$$

$$HOCl + NHCl_2 \rightarrow NCl_3 \text{ (trichloramine)} + H_2O$$

In the presence of excess chlorine, we obtain the total oxidation of the ammonia nitrogen into nitrogen gas:

$$2\,NH_3 + 3HClO \rightarrow N_2 + 3H_2O + 3H^+ + 3Cl^-$$

The competition between reactions depends on the wastewater pH, temperature, contact time and, above all, the initial chlorine/ammonia ratio. Monochloramine and dichloramine are the two dominant species in most cases. Chlorine reacts with certain nitrogenous organic matter to form organic chloramines. The sum of all chloramines is called available combined chlorine. The sum of the combined chlorine and free chlorine is called total chlorine.

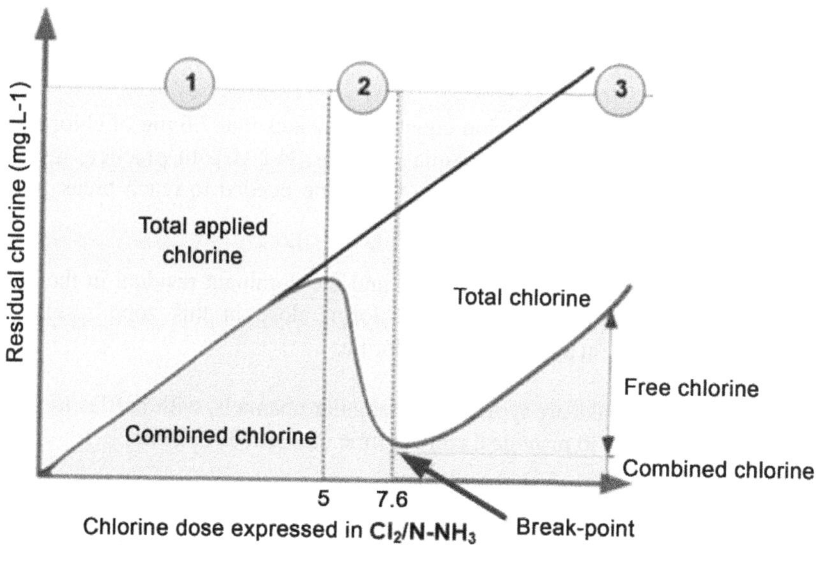

Figure 8.9. *Chlorination at break-point*

COMMENTS ON FIGURE 8.9.–

– *Zone 1: destruction of chlorine by mineral compounds.*

– *Zone 2: formation and then destruction of chloramines by adding more chlorine.*

– *Zone 3: production of active free chlorine.*

For water containing ammonia, disinfection should be performed by chlorination beyond the break-point. The break-point is determined in a jar test, by adding increasing amounts of chlorine to wastewater and then measuring the residual chlorine remaining after a given contact time (usually 30 min).

The shape of the curve of residual chlorine = f(added chlorine dose) is shown and commented in Figure 8.9. Several characteristic zones are identified. The dominant reaction in zone 1 is the reaction between chlorine and ammonia.

The maximum total chlorine theoretically occurs at a chlorine to ammonia ratio of 5:1 by weight (molar ratio of 1:1) with the formation of monochloramine. The addition of chlorine leads to an increase in the previous molar ratio and the formation of dichloramine begins. Dichloramine is unstable and adding more chlorine quickly leads to the formation of trichloramines. The additional dosages lead to complete oxidation of the ammonia nitrogen. The residual chlorine is at its lowest level. The break-point is reached.

The complete ammonia oxidation equation indicates that 7.6 mg of chlorine (Cl) is required to oxidize 1 mg of ammonia nitrogen ($N\text{-}NH_4$). In practice, given the chemical reactions competing, the dose of chlorine needed to reach break-point is about 9–10 mg/mg of ammonia nitrogen.

At this stage, chloramines are oxidized and the dominant residual in the zone 3 water is free chlorine. An increase in chlorine dose in this zone results in a corresponding increase in available free chlorine.

Chlorine contact tanks are typically rectangular channels, with baffles to prevent short-circuits, designed to provide a contact time of about 30–60 min.

However, to meet the stringent quality requirements, a chlorine contact time of up to 120 min is sometimes required for specific irrigation uses of recovered wastewater.

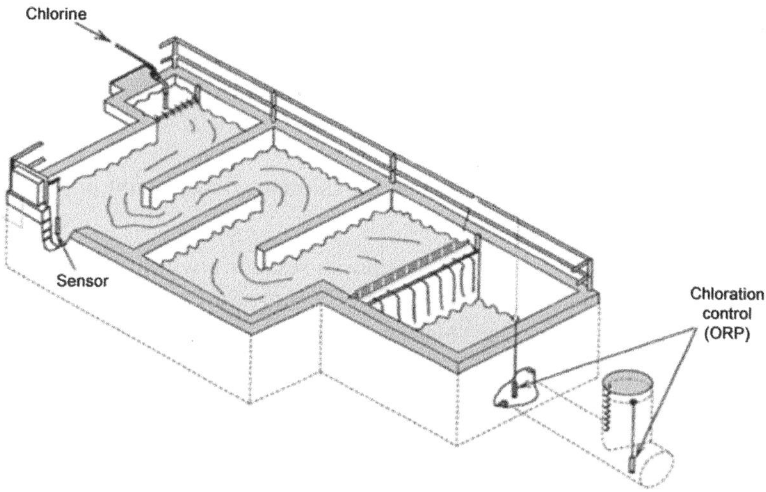

Figure 8.10. *Control of residual chlorine*

Figure 8.11. *Tertiary filtration and chlorination in the Ajman plant (United Arab Emirates)*

8.5.1.5. *Chlorination implementation*

Whether chlorine is introduced into the wastewater as gaseous chlorine or hypochlorite, only the pH and temperature of the medium determine the nature and proportion of the species present.

The most frequently used expressions for chlorine concentration in practice are as follows:

– the chlorometric degree (of a hypochlorite solution), which is the number of liters of chlorine gas (under normal temperature and pressure conditions) released by 1 L of hypochlorite solution. One chlorometric degree is equivalent to 3.17 $g \cdot L^{-1}$ of free chlorine;

– hypochlorite concentration is often expressed as free chlorine: 1 g of sodium hypochlorite (NaOCl) equals 0.96 g of free chlorine (Cl_2).

8.5.1.5.1. Liquid bleach

Hypochlorite is easier to implement. They are available as sodium or calcium salt. Sodium hypochlorite NaOCl is formed by dissolving chlorine gas in excess soda solution:

$$2NaOH + Cl_2 \rightleftharpoons 2NaOCl + H_2O$$

The commercial aqueous solution, commonly known as liquid bleach, is a greenish yellow liquid with pH 11. It is characterized by a chlorometric concentration of 47 to 50° corresponding to about 150 g/L of free chlorine. The solution is relatively unstable, forming chlorates and chlorides. Its concentration may decrease during storage depending on the exposure to light and temperature. This reagent is directly injected into the water to be treated usually by means of dosing pumps. For small or medium-sized plants, the use of bleach is less expensive (compared with chlorine gas), and equipment is usually reduced to storage and one or more dosing pumps.

The industrial liquid bleach can be delivered in 20–30 L bottles, 2,400 L containers or by tankers.

It must always be stored in a place that is protected from heat and light to slow down the decomposition process as much as possible. The amount of hypochlorite injected should be regularly monitored, as the strength of the storage solution may decrease over time.

Acids (such as hydrochloric acid or sulfuric acid) should never be stored in the same place as bleach to avoid any risk of contact between the two products (violent reaction with chlorine emissions). Liquid bleach must be stored in plastic (PVC, polyethylene, polyester resin, etc.) or ebonite steel tanks.

As a result of chlorine degassing, to avoid the deactivation of the dosing pump, vents must be provided on the suction and discharge lines.

As liquid bleach is an alkaline product, care must be taken to ensure that the service water used for its dilution is not hard, as there would be a risk of progressive clogging of the elements of the dosing pump, due to carbonate deposition. In such a situation, the following are required:

– either dilute liquid bleach with softened water;

– or add a sequestrant in the dilution water (e.g. sodium hexametaphosphate, at a rate of 50 g/100 L of water).

Liquid bleach is a corrosive and caustic product. Hypochlorite decomposes over time according to the temperature. The decomposition reactions are as follows:

$$2\ ClO^- \rightarrow ClO_2^- + Cl^-$$

$$ClO^- + ClO_2^- \rightarrow ClO_3^- + Cl^-$$

The half-life of hypochlorite depending on temperature is shown in Table 8.5.

Temperature (°C)	Half-life (j)
5	3,300
15	890
25	240
35	65
45	18

Table 8.5. *Decomposition of hypochlorite depending on temperature*

In the Middle East, this parameter is particularly monitored due to high temperatures.

8.5.1.5.2. Use of gas chlorine

For large plants, chlorine gas treatment is preferable (compared to liquid bleach), as it avoids the transport of water, which is always expensive, and because storage volumes are reduced.

Chlorine is a corrosive, irritating, greenish yellow gas, two and a half times heavier than air (d = 2.491). Its liquefaction temperature under 1 bar is –34°C. Its solubility in water at 20°C under pressure of 1 bar is 7–407 mg·L^{-1} of pure gas. Its redox potential is high (1.36 V at 25°C). Chlorine gas is sold in liquid form at 5–6 bar in 50 kg bottles or 1,000 kg tanks.

The distribution is ensured by means of chlorometers mounted directly on the containers and operating under vacuum, which allows to ensure an automatic closing of the container in case of leakage on the pipeline carrying the chlorine gas.

As chlorine is stored in liquid state and extracted in gaseous state, at a temperature of approximately 20°C, the maximum flows that can be extracted are 1 kg·h^{-1} for a 50 kg bottle and 20 kg·h^{-1} for a 1,000 kg tank.

Chlorine gas can also be used in medium-sized plants. The following equipment is required to implement disinfection using chlorine gas:

– the liquefied chlorine storage device: in 50 kg bottles in medium-sized plants, or in 1,000 kg tanks in larger plants;

– the chlorometer, ensuring the distribution of chlorine gas.

Chlorometers are mounted directly on the chlorine tank, or mounted on the wall, and operate under vacuum, which ensures automatic closing of the tank in case of leakage on the pipeline carrying the chlorine gas.

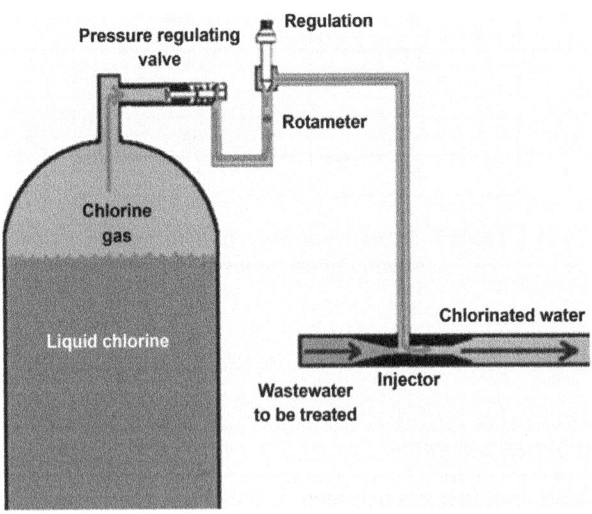

Figure 8.12. *Chlorination system with chlorine gas in bottle*

Since chlorine is stored in liquid form and extracted in gaseous state, and the maximum flows that can be extracted (at 15–20°C) are:

– 1–2 kg·h^{-1} for a bottle;

– 10 kg·h^{-1} for one tank.

For large plants, several tanks can be put into service simultaneously in parallel. If the flow of wastewater to be treated is very large and the number of chlorine tanks to be connected in parallel becomes prohibitive (>6), it is better to use a liquid chlorine evaporator, which can extract about 25% of the total tank capacity per hour. This system would include:

– one or more spirometer assemblies with flow control valve;

– one or more hydro ejectors, allowing the preparation of chlorine water by dissolving chlorine gas in service water.

The chlorine gas installation may also include the following equipment:

– an automatic inverter, to pass from the empty chlorine tank to the reserve tank without interrupting the chlorination (for plants where continuous monitoring cannot be ensured);

– a control assembly to link the chlorine flow rate to the wastewater to be treated and/or residual chlorine.

Chlorine is commercially available in liquid form, contained in bottles at a pressure of 15 bar and with an average capacity equal to 40 kg of chlorine. Pure dry chlorine is not corrosive, so containers are constructed of drawn, forged, seamless steel. Inside this casing, there is a gas/liquid equilibrium, and a space dedicated to the gas in equilibrium to compensate for pressure variations depending on the outside temperature, which must be between 10 and 40°C in storage. Chlorine storage is regulated as part of "Classified Facilities for Environmental Protection".

From the storage containers, chlorine is distributed to the dosing element called chlorometer, either in gaseous or liquid form. Direct sampling in the gas phase by expansion is limited due to endothermic evaporation. Direct warming of containers is prohibited for safety reasons. In this case, the ambient storage temperature maintained by air circulation around the containment enclosures determines the flow of the gas sampled. It reaches 10 kg·h^{-1} per 1,000 kg tank in a room at a temperature of 20°C. For larger quantities, the extraction of chlorine is carried out by sampling in liquid phase and forced evaporation. The evaporator consists of a bath with a thermostat at 70–80°C. The sampling capacity is increased by a factor of 5–10 in relation to the gas sampling system.

The transfer line downstream of the sampling system is thermostatically controlled at a temperature 3–5°C higher than the storage room temperature to prevent any risk of partial condensation. The chlorine gas is measured and expanded at the operating pressure in the chlorometer. The principle of this system is to dissolve chlorine gas using a hydro-ejector, which creates a negative pressure adjustable to the flow of chlorine. The capacity of chlorometers can reach 200 $kg \cdot h^{-1}$.

The concentrated aqueous solution from the chlorometer is mixed with the wastewater to be treated by an immersed system designed so that the ejection rate provides a good mixture quality for the intended application.

8.5.1.5.3. Electrochlorination

Electrochlorination is the production of sodium hypochlorite in situ by electrolysis of a sodium chloride (brine) solution.

The reactions are as follows:

– at the cathode: $Na^+ + e^- \rightarrow Na$;

– in water: $Na + 2H_2O \rightarrow 2\ NaOH + H_2$;

– at the anode: $2Cl^- \rightarrow Cl_2 + 2e^-$;

– in water: $Cl_2 + H_2O \rightleftarrows HCl + HClO$.

The formed hypochlorous acid HClO is neutralized by sodium hydroxide NaOH to give sodium hypochlorite.

An electrochlorination plant includes the following equipment:

– salt saturation: brine preparation;

– brine transfer pump to electrolyzers;

– softeners to produce softened water for the production of saturated brine and to dilute this brine at the electrolyzer (to avoid problems with scaling at the cathode);

– brine dilution pump;

– electrolysis cells, powered by a rectifier;

– storage tank for the produced hypochlorite equipped with a ventilation system to dilute and reject the hydrogen formed during electrolysis;

– dosing pump for hypochlorite dosage.

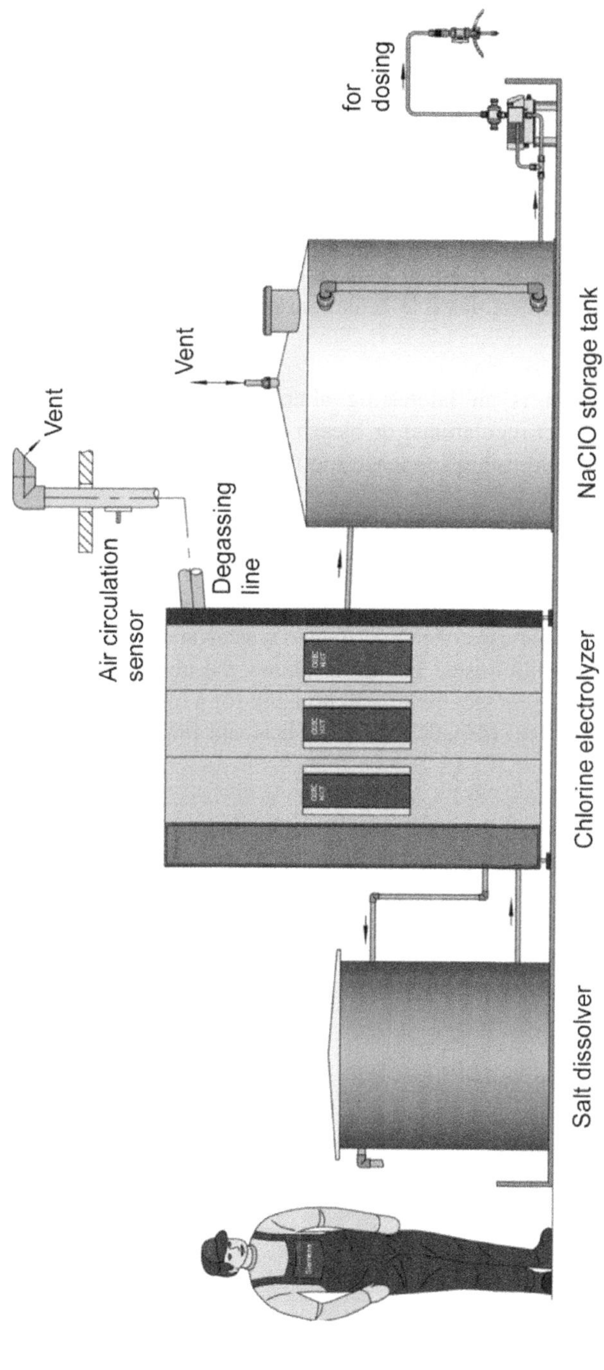

Figure 8.13. *Electrochloration plant*

The optimum performance of an electrochlorinator is at temperatures between 15 and 20°C. The declared energy consumption varies between 4.5 and 5.8 kW per kg Cl_2, and salt consumption varies between 3.1 and 3.5 kg NaCl per kg Cl_2. The salt used must be food grade.

The cost of producing sodium hypochlorite by electrolysis depends on the salt concentrations used and the desired hypochlorite concentration. The cost of producing softened water for brine production must be added to this. It is therefore necessary, at the industrial level, to find a compromise between the salt concentration used and the desired hypochlorite concentration, so as to produce hypochlorite under the best possible economic conditions (power consumption and salt consumption).

Electrochlorination is an interesting alternative to the use of chlorine gas (classified plants, strict regulations) or bleach (degradation over time of the stored industrial product due to its high concentration).

8.5.1.6. *Dosage and performance of chlorine disinfection*

The effectiveness of disinfection depends on Ct and contact time. The state of California requires a Ct of 450 $mg \cdot L^{-1} \cdot min^{-1}$ and a contact time of 90 min at maximum flow for some uses of reused treated wastewater. This condition allows a 4 log reduction of polioviruses. Table 8.6 shows the chlorine doses for different types of effluents, for a contact time of 30 min and for a total coliform concentration in raw wastewater of 10^7–10^9 UFC/100 mL (Haas and Engelbrecht 1980; Jacangelo et al. 1995).

The required dosage depends on the immediate demand for chlorine, the degradation of chlorine during contact time, and the residual chlorine concentration that is required. The immediate demand for chlorine depends on elements such as iron, manganese, nitrites and also BOD and residual COD. In the presence of N-NH_4, a demand for chlorine will be required for the formation of chloramines. The demand for residual free chlorine requires total neutralization of N-NH_4 up to break-point.

It should also be noted that the presence of SS protects some of the microorganisms that are attached to this matter. The decay rate of chlorine in the contact basin is between 0.03 and 0.07 $mg \cdot L^{-1} \cdot min^{-1}$.

Type of effluent		Dose of chlorine (mg·L^{-1})			
Fecal coliforms UFC/100 mL		1,000 UFC/100 mL	200 UFC/100 mL	23 UFC/100 mL	2.2 UFC/100 mL
Activated sludge	10^5–10^6	2–10	5–15	10–30	–
Nitrified effluent	10^5–10^6	4–10	5–15	8–18	–
Filtered effluent	10^4–10^6	4–8	5–15	6–20	8–30
Filtered nitrified effluent	10^4–10^6	4–8	5–15	6–16	8–20
Microfiltration	10–10^3	1–2	2–4	2–6	4–8
Reverse osmosis	–	–	–	–	0–2

Table 8.6. *Chlorine dosage guide depending on the effluent type and objectives*

The performance (expressed as free chlorine dosage) for viruses and *Giardia* is summarized in Tables 8.7 and 8.8.

They may however be different for specific microorganisms.

8.5.1.6.1. Viruses

	Disinfectant Ct (mg·L^{-1}·min^{-1}) with chlorine (pH 6.5–8.5)				
Virus log reduction	5°C	10°C	15°C	20°C	25°C
2	4	3	2	1	1
3	6	4	3	2	1
4	8	6	4	3	2

Table 8.7. *Ct (mg·L^{-1}·min^{-1}) required values for viruses*

8.5.1.6.2. Parasites

Giardia log inactivation	Ct ($mg \cdot L^{-1} \cdot min^{-1}$) with chlorine (pH 6.5–8.5)				
	Temperature (°C)				
	5	10	15	20	25
1.0 log	39–79	29–59	20–39	15–30	10–20
2.0 log	78–157	59–118	39–79	29–59	19–39
3.0 log	117–236	88–177	59–118	44–89	29–59

Table 8.8. *Ct ($mg \cdot L^{-1} \cdot min^{-1}$) required values for parasites*

Cryptosporidium are even more resistant protozoa than *Giardia lamblia*. The Ct values to be implemented are such that, in the treatment to make drinking water, it can be considered that chlorine is ineffective (Ct of 1,000 $mg \cdot min^{-1} \cdot L^{-1}$ at pH 6 to inactivate 1 log (90%)).

Protozoan (oo)cysts are more resistant to chlorine and almost no helminth eggs are inactivated by this disinfectant.

Thus, the values of residual chlorine, contact time and pH are the parameters that determine the effectiveness of chlorine disinfection. Chlorine is effective against bacteria and viruses, but its action will be more limited for protozoa, and it will have no effect on *Cryptosporidium* under normal treatment conditions.

It is essential to make a clear distinction between chlorine dose and residual chlorine. Confusion is often a source of errors that can result in ineffective disinfection. The chlorine dose is the amount of chlorine applied per liter of water to be treated. If the demand for chlorine of the wastewater is high (presence of ammonium ions, organic matter), the effective residual free chlorine will be minimal or even nil. It is necessary to reason in terms of dissolved residual free chlorine, the only one that is really effective, and which must be as stable as possible.

8.5.1.7. *Calculations of doses required to reduce total coliforms and coliforms for chlorine disinfection*

8.5.1.7.1. Chlorine disinfection (N-NH_4 < 2 $mg \cdot L^{-1}$)

Tables 8.9 and 8.10 provide calculations for the estimation of required doses based on the objectives for the disinfected water quality (Gyürék and Finch 1998; Liberti and Notarnicola 1999).

Parameters	Calculations	Units
Inlet average flow	A	$m^3 \cdot d^{-1}$
SS	B	$mg \cdot L^{-1}$
BOD5	C	$mg \cdot L^{-1}$
Temperature	D	°C
pH	E	–
Fe	F	$mg \cdot L^{-1}$
N-NO_2	G	$mg \cdot L^{-1}$
N-NH_4	H	$mg \cdot L^{-1}$
Total coliform inlet	I	$mg \cdot L^{-1}$
Fecal coliform inlet	J	MPN/100 mL
Required fecal coliform outlet	K	MPN/100 mL
Required total coliform outlet	L	MPN/100 mL
Log reduction required for total coliforms	log I-log L	–
Log reduction required for fecal coliforms	log J-log K	–
Contact time to obtain free chlorine	tc: 30	Min
Homogenization coefficient of contact basin	Bf: 0.8	–
	SS effects on residual microorganisms	
Fecal coliforms	M: 0.25 ($B^{1.96}$)	MPN/100 mL
Total coliforms	N: 0.9 ($B^{1.96}$)	MPN/100 mL
Enterococci	O: 0.1 ($B^{1.96}$)	MPN/100 mL
	Bacterial density coefficients in SS	
n total coliforms	2.8	–
b total coliforms	4	–
n′ fecal coliforms	2.8	–
b′ fecal coliforms	3	–
	Calculations of required chlorine concentrations	
Ct required for total coliforms	P: $b\ ((L - N)/(I - N))^{-1/n}$	$mg \cdot min^{-1} \cdot L^{-1}$
Ct required for fecal coliforms	Q: $b'\ ((K - M)/(J - M))^{-1/n'}$	$mg \cdot min^{-1} \cdot L^{-1}$
Required free chlorine	R	$mg \cdot L^{-1}$

Concentration based on total coliforms	S: P/tc	mg·L^{-1}
Concentration based on fecal coliforms	T: Q/tc	mg·L^{-1}
Chlorine consumption rate	U: 0.07	mg·min^{-1}·L^{-1}
Total Cl_2 required to reduce total coliforms	V: ((H*7.6) + R + S + (5*G) + (0.6*F) + (C*0.2) + (U*tc))/Bf	mg·L^{-1}
Total Cl_2 required to reduce fecal coliforms	W: ((H*7.6) + T + S + (5*G) + (0.6*F) + (C*0.2) + (U*tc))/Bf	mg·L^{-1}
Equivalent HOCl required for total coliforms	X: 52.2*V/71	mg·L^{-1}
Equivalent HOCl required for fecal coliforms	Y: 52.2*W/71	mg·L^{-1}
Average consumption of HOCl/d to reduce fecal coliforms	Z: Y*A/1,000	kg·j^{-1}
	Summary	
Free chlorine	R	mg·L^{-1}
Demand for Cl_2 to oxidize NH_4^+	AB: 7.6*H	mg·L^{-1}
Injected Cl_2	W	mg·L^{-1}
$Cl_2 + H_2O \leftrightarrow HOCl + H^+ + Cl^-$ Equilibrium factor K_H at 25°C	0.00045	(mole·L^{-1})^2
Available free chlorine [HOCl] + [OCl$^-$]	AC: R*(–(AB/2) + S)	mg·L^{-1}
HOCL $\leftrightarrow$ H^+ + OCl^- HOCL ionization constant ki at 20°C	0.000000026	mole·L^{-1}

Table 8.9. *Determination of chlorine concentration for disinfection (N-NH_4 < 2 mg·L^{-1})*

8.5.1.7.2. Example of calculation for chlorine disinfection ($N\text{-}NH_4 < 2\ mg{\cdot}L^{-1}$)

Table 8.10 shows an example of chlorine disinfection calculation for wastewater containing $N\text{-}NH_4 < 2\ mg{\cdot}L^{-1}$.

Parameters	**Calculations**	**Units**
Inlet average flow	50,000	$m^3{\cdot}d^{-1}$
Inlet maximum flow	–	$m^3{\cdot}d^{-1}$
SS	4.0	$mg{\cdot}L^{-1}$
BOD5	5.0	$mg{\cdot}L^{-1}$
Temperature	25	°C
pH	8	–
Fe	0.0	$mg{\cdot}L^{-1}$
$N\text{-}NO_2$	0.00	$mg{\cdot}L^{-1}$
$N\text{-}NH_4$	1.000	$mg{\cdot}L^{-1}$
Total coliform inlet	1.00E+06	MPN/100 mL
Fecal coliform inlet	1.00E+05	MPN/100 mL
Required fecal coliform outlet	23	MPN/100 mL
Required total coliform outlet	2.00E+02	MPN/100 mL
Log reduction required for total coliforms	3.70E+00	–
Log reduction required for fecal coliforms	3.64E+00	–
Contact time to obtain free chlorine	30	min
Homogenization coefficient of the contact basin	0.8	–
SS effects on residual microorganisms		
Fecal coliforms	4	MPN/100 mL
Total coliforms	13.6	MPN/100 mL
Enterococci	1.61	MPN/100 mL
Bacterial density coefficients in SS		
n total coliforms	2.8	–
b total coliforms	4	–
n′ fecal coliforms	2.8	–
b′ fecal coliforms	3	–

Calculations of required concentrations		
Ct required for total coliforms	85.91	$mg \cdot min^{-1} \cdot L^{-1}$
Ct required for fecal coliforms	63.74	$mg \cdot min^{-1} \cdot L^{-1}$
Free chlorine required	1.00	$mg \cdot L^{-1}$
Concentration based on total coliforms	2.86	$mg \cdot L^{-1}$
Concentration based on fecal coliforms	2.12	$mg \cdot L^{-1}$
Chlorine consumption	0.083	$mg \cdot min^{-1} \cdot L^{-1}$
Total Cl_2 required to reduce total coliforms	15.6	$mg \cdot L^{-1}$
Total Cl_2 required to reduce fecal coliforms	14.7	$mg \cdot L^{-1}$
Equivalent HOCl required for total coliforms	11.5	$mg \cdot L^{-1}$
Equivalent HOCl required for fecal coliforms	10.8	$mg \cdot L^{-1}$
Average consumption of HOCl/d to reduce fecal coliforms	538.7	$kg \cdot j^{-1}$

Table 8.10. *Example of calculation of chlorine concentration for disinfection (N-NH_4 < 2 $mg \cdot L^{-1}$)*

8.5.1.7.2. Quick estimate

– No: fecal coliform inlet, 1.0×10^5 UFC/100 mL.

– Nt: outlet, 23 UFC/100 mL.

– Desired log reduction: log 1.0 10^5 – log 23 = 3.63 log.

– Nitrified filtered effluent.

– $\frac{Nt}{No} = (1 + 0.23\ C_R * t)^{-3}$

- C_R: residual chlorine in $mg \cdot L^{-1}$;

- t = contact time in min.

– Calculations:

- $\frac{23}{1.0 \times 10^5} = (1 + 0.23\ C_R * t)^{-3}$

- $23 \times 10^{-5} = (1 + 0.23\ C_R * t)^{1/3}$

- $1 + 0.23\ C_R * t = 16.27$

- $C_R * t = 66.39$

- contact time = 60 min;

- $CR = \frac{66.39}{60} = 1.10\ mg \cdot L^{-1}$ in residual total chlorine;

- for 4 log reduction, 8 $mg \cdot L^{-1}$ + 1.10 $mg \cdot L^{-1}$ of residual total chlorine must be injected, giving a total of 9.1 $mg \cdot L^{-1}$ of chlorine to be injected.

This calculation does not take into account the immediate demand for chlorine (BOD5, metals, nitrites, color) nor the presence of SS and ammonia.

8.5.2. *Chlorine dioxide disinfection*

Chlorine dioxide is a compound of chlorine at oxidation level 4, with the ClO_2 formula and 67.45 g molar mass. It is a relatively strong oxidant, but four-fifths of its oxidizing potential can only be used at acidic pH. At pH values close to neutrality, the chlorine dioxide transformation will be mainly in the form of chlorite according to the first reaction:

$$ClO_2 + e^- \rightarrow ClO_2^-\ E°: 0.95\ V \text{ at } 25°C$$

$$ClO_2^- + 4H_3O^+ + 4e^- \rightarrow Cl^- + 6H_2O$$

Chlorine dioxide is highly soluble in water and its solubility depends on temperature and pressure (70 $g.L^{-1}$ at 20°C and 1 atm). Its dissolution in water at neutral pH results in a mixture of chlorous and chlorine acids:

$$2ClO_2 + H_2O \rightleftharpoons HClO_2 + HClO_3$$

In alkaline medium, chlorine dioxide undergoes an irreversible dismutation into chlorite and chlorate ions. The reaction is irreversible at pH higher than 11:

$$2ClO_2 + 2OH^- \rightleftharpoons ClO_2^- + ClO_3^- + H_2O$$

In neutral or acidic medium, chlorine dioxide reacts slowly to form the chloric acid:

$$6ClO_2 + 3H_2O \rightleftharpoons 5HClO_3 + HCl$$

Chlorine dioxide is an orange, unstable gas. It explodes under the effect of heat, exposure to light or contact with organic matter. Therefore, it is prepared in solution at the site of use.

8.5.2.1. *Preparation of chlorine dioxide*

Chlorine dioxide, ClO_2, is an unstable product that dissociates very easily into gas phase. It is explosive when present in air at levels above 10%, but harmless when dissolved in water.

To avoid any risk related to transport and storage, chlorine dioxide is prepared at the site of its use (in situ) and must be used immediately.

Chlorine dioxide is prepared by chemical reaction between chlorine and sodium chlorite. This reaction takes place in the liquid phase and can be conducted in two ways:

1) Action of elemental chlorine on sodium chlorite:

$$2NaClO_2 + Cl_2 \rightarrow 2ClO_2 + NaCl$$

For the manufacture of chlorine dioxide from chlorine gas, in modern chlorinated water enrichment loop reactors, 1 g of chlorine dioxide is obtained with:

- 0.6 g of chlorine;
- 1.34 g of sodium chlorite.

2) Action of hypochlorite in the presence of hydrochloric acid:

$$NaOCl + HCl \rightarrow Cl_2 + NaCl + H_2O$$

$$\downarrow$$

$$Cl_2 + 2\,NaClO_2 \rightarrow 2\,ClO_2 + NaCl$$

The second reaction channel requires the use of three reagents (NaOCl, HCl and $NaClO_2$), but has the advantage of not involving chlorine gas, as is the case for channel 1.

The control of reaction stoichiometry is very important to minimize possible by-products:

– If the chlorine concentration is insufficient, a mixture of chlorine and chlorites is obtained as the reaction is incomplete. Although there is no regulation yet on chlorite content in drinking water, it is advisable to maintain a level below

0.3–0.4 mg/L. Although their role is not well established yet, chlorites are suspected of having an effect on the human body similar to that of nitrites (reduction of hemoglobin, formation of carcinogens by reaction with certain amines contained in medicines, etc.). To introduce a sufficient amount of chlorine (from chlorine gas) into the ClO_2 preparation reactor, commercially available devices work with a chlorine enrichment loop.

– If chlorine is dosed in excess, a mixture of chlorine and chlorine dioxide is obtained, and this excess chlorine can give rise to haloforms (THM) or sapid compounds.

Chlorine dioxide can also be obtained by the action of a strong acid (hydrochloric, sulfuric) on sodium chlorite according to:

$$4HCl + 5NaClO_2 \rightarrow 4ClO_2 + 2H_2O + 5NaCl \text{ (for the hydrochloric acid)}$$

Note that 1 $mg{\cdot}L^{-1}$ of ClO_2 equates to 2.63 $mg{\cdot}L^{-1}$ of Cl_2 in oxidizing power.

This preparation method may give rise to various secondary reactions, with the formation of undesirable products such as chlorates. This manufacturing method also consumes 25% more $NaClO_2$ than the reaction with chlorine for the same amount of chlorine dioxide produced.

Sodium chlorite used in the manufacture of chlorine dioxide is usually delivered as a 25% solution (density 1.21–1.22), in cylinders, containers or by tankers. It must be stored in a tank made of an inert material (PVC, polyethylene, etc.).

Because of its oxidizing nature, sodium chlorite must be stored away from heat, moisture and light in a ventilated room reserved exclusively for the storage of this product. Chlorine dioxide has a faster and more persistent disinfecting action than chlorine. Since chlorine dioxide is explosive at concentrations greater than 10% by volume in air, it is always produced in situ, preferably and for safety by the reaction of sodium chlorite with chlorine gas or in hydrochloric acid medium.

Soda chlorite can be delivered either in powder for the preparation of a solution at a maximum concentration of up 200 $g{\cdot}L^{-1}$, but it is then necessary to use softened water to avoid precipitation of carbonates, or in concentrated solution at 300 $g{\cdot}L^{-1}$ and 1.23 density. It is a hazardous product to handle, and concentrated solutions should be kept at a temperature above 10°C. The reagents and reactor (generator) are installed in separate rooms opening directly to the outside.

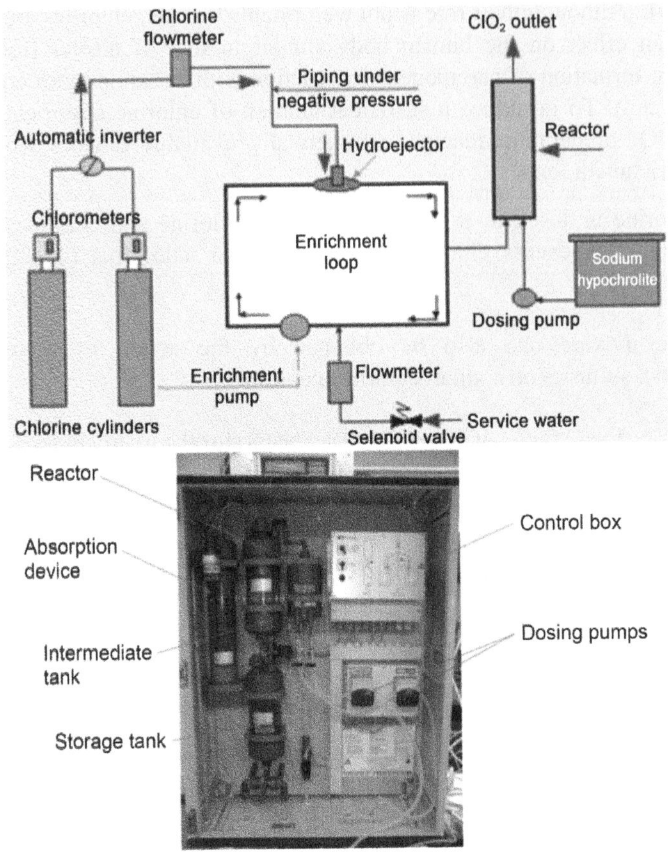

Figure 8.14. *Example of installation to prepare chlorine dioxide. For a color version of this figure, see www.iste.co.uk/gaid/reuse1.zip*

A chlorine dioxide preparation and dosing installation consists mainly of:

– a storage for chlorine gas;

– a storage for concentrated sodium chlorite;

– a chlorine dioxide generator;

– a chlorine dioxide transfer tank for dosing with dosing pumps.

The generator is operated by a level control system placed in the transfer tank. The room for chlorine dioxide preparation is a specific construction with natural ventilation. The storage of chlorine cylinders is in a building attached to the dioxide and equipped with a mesh opening.

After dilution in the transfer tank, the dioxide solution is taken up by two dosing pumps (one of which is a backup). The dosing pumps are equipped with frequency inverters to modulate the rate of injected reagent.

The standard treatment rates are 0.3–0.6 g ClO_2/m^3 of treated wastewater. The autonomy required by the specifications, spread over a period of 1–2 months, is obtained using chlorine cylinders (equipped with an automatic inverter) and calculated based on an average processing rate of 0.4 $g{\cdot}m^{-3}$.

In addition, as a safety measure, the chlorine dioxide production unit is equipped with gas detection, a sensor for chlorine gas and a flashing light.

The instrumentation usually provided at this workstation consists of a pH meter for monitoring the correct operation of the generator. Chlorine production has a flowmeter (measurement of the chlorine flow rate), capacitive sensors (triggering/stopping the generator), threshold levels on the chlorite tank, an in-line analyzer (residual chlorine and regulation of chlorine injection) and level switches (level detectors used as indicators for the operator).

8.5.2.2. *Performance of chlorine dioxide*

ClO_2 is a bactericide equal to or superior to chlorine in disinfectant. ClO_2 has been shown to be an effective virucidal agent, more effective than chlorine in inactivating viruses. One possible explanation is that ClO_2 is adsorbed by a protein and viruses have a protein shell, and the adsorption of ClO_2 on this shell could cause the inactivation of the virus.

ClO_2 has an extremely high oxidizing potential, which is probably why it has such strong germicidal powers. Possible bactericidal mechanisms may include the inactivation of critical enzyme systems or disruption of protein synthesis (Li et al. 2001).

It was observed that, unlike chlorine (whose different pH-dependent forms have very different disinfectant power), the effects of chlorine dioxide on bacteria, viruses and protozoa are not altered by changes in pH (in the range 6–9). The kinetics of chlorine dioxide inactivation of microorganisms generally follow Watson's law ($\log N_t/N_o = -k\, C^n\, t^1$). Depending on the nature of the microorganisms, the value of the coefficient *n* varies. Thus, it is equal to 2.5 for *Pseudomonas fluorescens*, 2 for *E. coli*, 0.55 for *Naegleria gruberi* and 1 for *Giardia* and *Cryptosporidium*.

As with other disinfectants, the effectiveness of chlorine dioxide depends on temperature. Tables 8.11–8.13 show the different efficiencies of chlorine dioxide for different microorganisms.

Type of effluent		Dose of ClO_2 ($mg \cdot L^{-1}$) Contact time: 10–15 min			
Total coliforms UFC/100 mL		1,000 UFC/100 mL	200 UFC/100 mL	23 UFC/100 mL	2.2 UFC/100 mL
Activated sludge	10^5–10^6	1.0–1.5	1.5–2	4–6	5–7
Filtered effluent	10^4–10^6	0.8–1.2	1.3–1.8	2.5–4	4–5
Filtered effluent + GAC	10^3–10^5	0.6–1.0	1.2–1.5	2–3	3–4

Table 8.11. *Chlorine dioxide dosing guide depending on the type of effluent and objectives*

8.5.2.2.1. Viruses

	Disinfectant Ct ($mg \cdot L^{-1} \cdot min^{-1}$) with chlorine dioxide (pH 6.5–8.5)				
Virus log reduction	5°C	10°C	15°C	20°C	25°C
2	5.6	4.2	2.8	2.1	1.4
3	17.1	12.8	8.6	6.4	4.3
4	33.4	25.1	16.7	12.5	8.4

Table 8.12. *Ct ($mg \cdot L^{-1} \cdot min^{-1}$) required values for viruses*

8.5.2.2.2. Parasites

The data in these tables show that it is possible to perform a post-disinfection with chlorine dioxide, of course after defining the objectives of the disinfection (especially the bactericidal action) while ensuring that the chlorite concentration does not exceed 0.2 $mg \cdot L^{-1}$ at the consumer tap.

Giardia inactivation log	Temperature (°C)					
	≤1	5	10	15	20	25
0.5 log	10.0	4.3	4.0	3.2	2.5	2.0
1.0 log	21.0	8.7	7.7	6.3	5.0	3.7
1.5 log	32.0	13.0	12.0	10.0	7.5	5.5
2.0 log	42.0	17.0	15.0	13.0	10.0	7.3
2.5 log	52.0	22.0	19.0	16.0	13.0	9.0
3.0 log	63.0	26.0	23.0	19.0	15.0	11.0

Table 8.13. *Ct values for the inactivation of* Giardia lamblia *cysts with chlorine dioxide at a pH between 6 and 9*

8.5.2.3. *Advantages and disadvantages of chlorine dioxide in wastewater*

Advantages	Disadvantages	Advantage compared to chlorine
The advantages of chlorine dioxide can be summed up as follows: – high germicidal power pH independent with limits from 6 to 10; – does not have any taste; – does not create trihalomethanes; – persistence of its disinfectant power in the network, double that of chlorine.	No action on ammonia nitrogen Formation of chlorites with the presence of organic matter	No formation of chloramines of organic derivatives → lower dosage Its germicidal power remains important with a high pH Slower reaction time (15–20 min) The residual disappears quickly More efficient against viruses

Table 8.14. *Advantages and disadvantages of chloride dioxide*

8.5.2.4. *Calculations of the doses required to reduce total coliforms and fecal coliforms for disinfection with chlorine dioxide*

8.5.2.4.1. Theoretical calculations for ClO_2 disinfection

Table 8.15 provides a calculation method for the disinfection of wastewater using ClO_2. As explained above, this oxidant does not react with ammonia nitrogen, so this calculation does not take into account the concentration of this element in the water to be treated.

Parameters	Calculations	Units
Average inlet flow	A	$m^3 \cdot d^{-1}$
SS	B	$mg \cdot L^{-1}$
BOD5	C	$mg \cdot L^{-1}$
Temperature	D	°C
pH	E	–
Fe	F	$mg \cdot L^{-1}$
N-NO$_2$	G	$mg \cdot L^{-1}$
N-NH$_4$	H	$mg \cdot L^{-1}$
Total coliform inlet	I	$mg \cdot L^{-1}$
Fecal coliform outlet	J	MPN/100 mL
Required fecal coliform outlet	K	MPN/100 mL
Required total coliform outlet	L	MPN/100 mL
Log reduction required for total coliforms	log I-log L	–
Log reduction required for fecal coliforms	log J-log K	–
Contact time to obtain free chlorine	tc: 30	Min
Homogenization coefficient of the contact basin	Bf: 0.8	–
SS effects of residual microorganisms		
Fecal coliforms	M: 0.25 ($B^{1.96}$)	MPN/100 mL
Total coliforms	N: 0.9 ($B^{1.96}$)	MPN/100 mL
Enterococci	O: 0.1 ($B^{1.96}$)	MPN/100 mL

Parameters	Calculations	Units
Bacterial density coefficients in SS		
n total coliforms	2.18	–
b total coliforms	0.89	–
n′ fecal coliforms	2.18	–
b′ fecal coliforms	0.89	–
Calculations of required ClO_2 concentrations		
Ct required for total coliforms	P: $b\ ((L - N)/(I - N)))^{-1/n}$	$mg \cdot min^{-1} \cdot L^{-1}$
Ct required for fecal coliforms	Q: $b'\ ((K - M)/(J - M))^{-1/n'}$	$mg \cdot min^{-1} \cdot L^{-1}$
Required residual ClO_2	R	$mg \cdot L^{-1}$
Concentration based on total coliforms	S: P/tc	$mg \cdot L^{-1}$
Concentration based on fecal coliforms	T: Q/tc	$mg \cdot L^{-1}$
ClO_2 consumption rate	U: 0.002	$mg \cdot min^{-1} \cdot L^{-1}$
Required total ClO_2 to reduce total coliforms	V: (R + S + (1.47*G) + (1.2*F) + (0.2*C) + (U*tc))/Bf	$mg \cdot L^{-1}$
Required total ClO_2 to reduce fecal coliforms	W: (T + S + (1.47*G) + (1.2*F) + (0.2*C) + (U*tc))/Bf	$mg \cdot L^{-1}$
Average ClO_2/d consumption to reduce fecal coliforms	Z: W*A/1,000	$kg \cdot j^{-1}$

Table 8.15. *ClO_2 concentration determination for disinfection*

8.5.2.4.21. Example of calculation for ClO_2 disinfection

Table 8.16 shows an example calculation for the disinfection of wastewater using ClO_2.

Parameters	Calculations	Units
Inlet average flow	50,000	$m^3 \cdot d^{-1}$
Inlet maximum flow	–	$m^3 \cdot d^{-1}$
SS	5.0	$mg \cdot L^{-1}$
BOD5	5.0	$mg \cdot L^{-1}$
Temperature	20	°C
pH	8.0	–
Fe	0.0	$mg \cdot L^{-1}$
$N\text{-}NO_2$	0.00	$mg \cdot L^{-1}$
$N\text{-}NH_4$	5.000	$mg \cdot L^{-1}$
Total coliform inlet	1.00E+05	$mg \cdot L^{-1}$
Fecal coliform inlet	1.00E+04	MPN/100 mL
Required fecal coliform outlet	20	MPN/100 mL
Required total coliform outlet	1.00E+02	MPN/100 mL
Log reduction required for total coliforms	2.70E+00	–
Log reduction required for fecal coliforms	3.00E+00	–
Contact time to obtain free chlorine	30	Min
Homogenization coefficient of contact basin	0.8	–
SS effects on residual microorganisms		
Fecal coliforms	6	MPN/100 mL
Total coliforms	21.10	MPN/100 mL
Enterococci	2.34	MPN/100 mL
Bacterial density coefficients in SS		
n total coliforms	2.18	–
b total coliforms	0.89	–
n′ fecal coliforms	2.18	–
b′ fecal coliforms	0.89	–
Calculations of required ClO_2 concentrations		
Ct required for total coliforms	23.59	$mg \cdot min^{-1} \cdot L^{-1}$
Ct required for fecal coliforms	18.05	$mg \cdot min^{-1} \cdot L^{-1}$
Required residual ClO_2	1.00	$mg \cdot L^{-1}$
Concentration based on total coliforms	0.79	$mg \cdot L^{-1}$
Concentration based on fecal coliforms	0.60	$mg \cdot L^{-1}$
ClO_2 consumption rate	0.0400	$mg \cdot min^{-1} \cdot L^{-1}$
Total ClO_2 required to reduce total coliforms	5.0	$mg \cdot L^{-1}$
Total ClO_2 required to reduce fecal coliforms	4.8	$mg \cdot L^{-1}$
Average ClO_2/d consumption to reduce fecal coliforms	249.1	$kg \cdot j^{-1}$
Residual $N\text{-}NH_4$	5.000	$mg \cdot L^{-1}$

Table 8.16. *Example of calculations of ClO_2 concentration for wastewater disinfection*

8.5.3. *Chloramination*

8.5.3.1. *Principle*

When chlorine and ammonia are introduced into water:

– Chlorine hydrolyzes rapidly to hypochlorous acid in water at pH 6.5–8.5 and 10–25°C. At pH 7.5 and 25°C, equal concentrations of hypochlorous acid (HOCl) and hypochlorite (OCl^-) exist.

– Ammonia dissolves easily in water and the form it takes depends heavily on pH and temperature. The proportion of non-ionized ammonia increases when pH exceeds 7.5. Equal concentrations of NH_3 and NH_4^+ exist at approximately pH 9.3 and 25°C.

Thus, when chlorine (or hypochlorite) is introduced into water containing ammonia, chloramines are formed, which constitute a fraction of the "combined chlorine" (as the remainder is organic chloramines). The nitrogen organic compounds obtained are monochloramine (NH_2Cl), dichloramine ($NHCl_2$) and trichloramine (NCl_3).

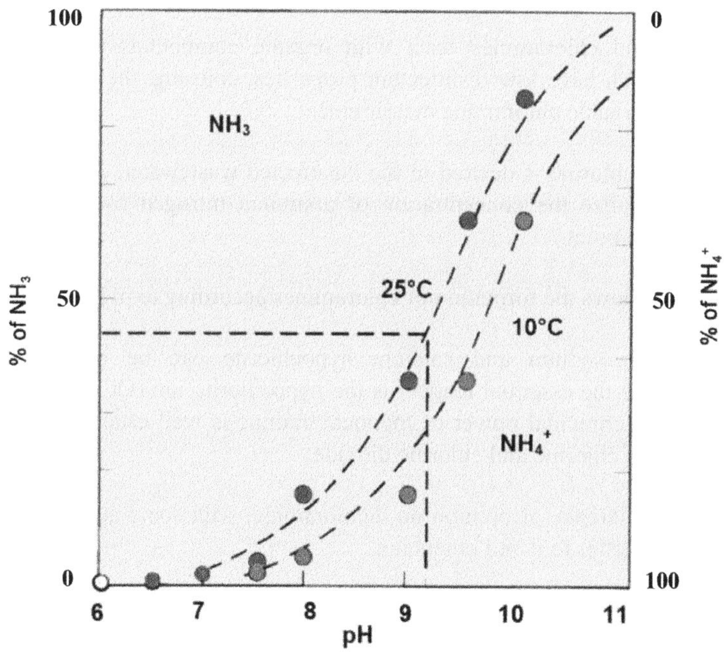

Figure 8.15. *Ionized and non-ionized forms of ammonia*

Formation reactions of inorganic chloramines are as follows:

$NH_3 + HClO \rightarrow NH_2Cl$ (monochloramine) $+ H_2O$

$NH_2Cl + HClO \rightarrow NHCl_2$ (dichloramine) $+ H_2O$

$NHCl_2 + HClO \rightarrow NHC_3$ (trichloramine) $+ H_2O$

– The proportion of mono-, di- and trichloramine formed depends on a number of factors such as pH, temperature, alkalinity, the chlorine dose introduced and the $Cl_2/N\text{-}NH_3$ ratio.

– Monochloramine (NH_2Cl) is normally the most sought-after chloramine species due to its biocidal properties and relative stability and because it causes few taste and odor problems.

– Optimal conditions for monochloramine formation are pH 8.3, 25°C and a weight ratio of 4/1–5/1 of $Cl_2/N\text{-}NH_3$.

– Normally, at pH of 6.5–9.0 and 10–25°C, monochloramine forms in seconds.

– Chlorine and chloramines react with organic compounds to form organic chloramines, which have low disinfectant properties, consume the disinfectant and interfere with inorganic chloramine measurement.

– When free chlorine is desired in the disinfected wastewater, it is important to completely neutralize the concentration of ammonia nitrogen by applying doses higher than break-point.

Figure 8.16 shows the formation of chloramines according to pH.

Chlorine gas, sodium and calcium hypochlorite can be used to prepare chloramines since the essential reagent is the hypochlorite ion (OCl^-). As shown in Table 8.17, the germicidal power of monochloramine is well established, although lower than that of chlorine and chlorine dioxide.

There is a difference of opinion on dichloramine, with some authors attributing disinfectant properties to it and others not.

The major advantage of chloramines is to enable disinfection without generating halogenated by-products with organic matter (THM) and have a significant residual power (higher than chlorine).

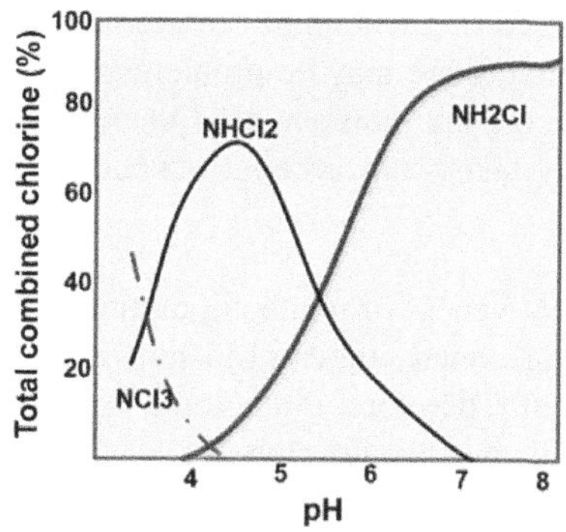

Figure 8.16. *Formation of chloramines depending on pH*

Disinfectant	E° (V) at 25°C
Monochloramine (NH_2Cl)	0.75
Dichloramine ($NHCl_2$)	0.79

Table 8.17. *Oxidation-reduction potential of chloramines compared to other disinfectants*

8.5.3.2. *Implementation*

Chloramination involves introducing both ammonia (or ammonia water) and chlorine (chlorine gas or hypochlorite) into the wastewater to be treated under controlled pH and stoichiometric reaction conditions.

A Cl/N-NH_3 ratio of between 3 and 5 (mg Cl·mg^{-1} N-NH_3) and a pH greater than or equal to 7 is generally recommended to promote the formation of monochloramine, whose germicidal power has been recognized.

Unlike chlorine, chlorine dioxide or ozone disinfection cases, there are no standard Ct recommendations for monochloramine.

To avoid the formation of THM, it is preferable to inject ammonia in pre-chlorinated water after a contact time between water and chlorine of at least 10 min. However, when the THM risk arises, it is better to dose ammonia first. The usual NH_2Cl dosages are between 2.0 and 4.0 mg·L^{-1}. The residual admitted in distribution networks is 0.5 mg·L^{-1}.

Injection points are based on the addition of ammonia (NH_3) and then chlorine. When chlorine is added first, there may be problems with taste and odor due to the reaction of chlorine with organic compounds (phenolic compounds, toxins, etc.). This consideration is only taken into account depending on the use of the reused treated wastewater.

As the germicidal effectiveness of monochloramine is 200 times lower than that of free chlorine, very long contact times (120 mins) are required for disinfection performance to be achieved. Therefore, if ammonia is added first, a means to ensure that Ct requirements are met must be developed.

When ammonia (<0.5 $mg \cdot L^{-1}$) remains in the filtered water, two solutions exist to implement chloramination:

– perform a break-point to eliminate all residual ammonia, and then implement chlorination on controlled Cl_2/N ratios;

– perform chloramination taking into account the ammonia present. However, as this value varies greatly during the year, it is important to monitor it (ammonia meter) precisely in order to avoid chlorine deficits.

8.5.3.3. *Performance of monochloramine*

Inactivation of *E. coli* by monochloramine shows that it reacts readily with four amino acids, cysteine, cystine, methionine and tryptophan. The inactivation mechanism involves inhibition of proteins or protein-induced processes such as respiration. As with free chlorine, the virus inactivation mechanism depends on factors such as the type of virus and the disinfectant concentration.

8.5.3.3.1. Bacteria

	Disinfectant Ct ($mg \cdot L^{-1} \cdot min^{-1}$) with monochloramine (pH 7)
Bacteria log reduction	**20°C**
1	4–6
2	12–20
3	30–75
4	200–250

Table 8.18. *Ct ($mg \cdot L^{-1} \cdot min^{-1}$) required values for viruses*

8.5.3.3.2. Viruses

	Disinfectant Ct ($mg \cdot L^{-1} \cdot min^{-1}$) with monochloramine (pH 6.5–8.5)				
Virus log reduction	**5°C**	**10°C**	**15°C**	**20°C**	**25°C**
2	857	643	428	321	214
3	1,423	1,067	712	534	356
4	1,988	1,491	994	746	497

Table 8.19. *Ct ($mg \cdot L^{-1} \cdot min^{-1}$) required values for viruses*

***Giardia* log inactivation**	**Temperature (°C)**				
	5	**10**	**15**	**20**	**25**
0.5	365	310	250	185	125
1	735	615	500	370	250
1.5	1,100	930	750	550	375
2	1,470	1,230	1,000	735	500
2.5	1,830	1,540	1,250	915	625
3	2,200	1,850	1,500	1,100	750

Table 8.20. *Inactivation Ct ($mg \cdot L^{-1} \cdot min^{-1}$) of* Giardia *with monochloramine*

Chloramines are less potent disinfectants for viruses and protozoa. It is therefore difficult to meet performance targets for protozoa such as *Cryptosporidium* and *Giardia*, because very high Ct are required. It takes between 2 and 8 h of reaction time with 1 $mg \cdot L^{-1}$ of residual NH_2Cl to achieve 2 log inactivation of poliovirus, Coxsackie virus B1 and B5, whereas only 0.25 $mg \cdot L^{-1}$ of free chlorine and a reaction time of 15 min with chlorine to obtain the same inactivation. Similarly, 8 $mg \cdot L^{-1}$ of NH_2Cl is required to obtain 2 log inactivation of *Entamoeba histolytica* cysts, while only 3.8 mg·L^{-1} of free chlorine is needed for an equivalent reaction time of 10 min.

8.5.3.4. *Advantages and disadvantages of chloramination in wastewater*

The advantages and disadvantages of chloramination are shown in Table 8.21.

Advantages	Disadvantages
The advantages of chloramination can be summed up as follows: – pH independent germicidal power with limits from 6 to 9; – longer residual time than chlorine and chlorine dioxide; – persistence of its disinfectant power in the network double of that of chlorine; – chloramines are not as reactive with organic matter as free chlorine in the formation of THMs; – residual monochloramine is stabler and more durable than free chlorine offering better protection against bacterial regrowth in distribution networks; – chloramines do not react to organic compounds; – no formation of bromate; – no formation of trihalomethanes; – no formation chlorophenols.	– Lower inactivation for viruses and protozoa than chlorine, ozone and chlorine dioxide; – cannot be applied in all cases depending on microorganisms to be eliminated and the required performance.

Table 8.21. *Advantages and disadvantages of chloramination*

8.5.3.5. *Calculations of doses required to reduce total coliforms and fecal coliforms for disinfection with chloramines*

8.5.3.5.1. Theoretical calculations of chloramination of wastewater

Table 8.22 provides a calculation method for the disinfection of wastewater with chloramines.

Parameters	Calculations	Units
Inlet average flow	A	$m^3 \cdot h^{-1}$
SS	B	$mg \cdot L^{-1}$
Temperature	C	°C
pH	D	
BOD5	E	$mg \cdot L^{-1}$
Fe	F	$mg \cdot L^{-1}$
N-NO_2	G	$mg \cdot L^{-1}$
N-NH_3	H	$mg \cdot L^{-1}$
Total coliform inlet	I	MPN/100 mL
Fecal coliform inlet	J	MPN/100 mL
Required fecal coliform outlet	K	MPN/100 mL
Required total coliform outlet	L	MPN/100 mL
Log reduction required for total coliforms	M: log I-log K	–
Log reduction required for fecal coliforms	N: log J-log L	–
Contact time to obtain free chlorine	tc	min
Homogenization coefficient of contact basin	O	–
SS effects on microorganisms		
Fecal coliforms	P: $0.25(B^{1.96})$	MPN/100 mL
Total coliforms	Q: $0.15(B^{1.96})$	MPN/100 mL
Enterococci	R: $0.10(B^{1.96})$	MPN/100 mL
Concentration in NH_3 and NH_4		
NH_3	17H/14	mg/L
NH_4	18H/14	mg/L
Bacterial density concentration on SS		
n total coliforms	2.18	–
b total coliforms	0.89	–
n′ fecal coliforms	2.18	–
b′ fecal coliforms	0.89	–

Parameters	Calculations	Units
Calculation of concentrations in chlorine and chloramines		
Ct required for NH_2Cl for total coliforms	S: $b^*((K - Q)/(I - Q))^{-1/n}$	$mg \cdot NH_2Cl \cdot L^{-1} \cdot min^{-1}$
Ct required for NH_2Cl for fecal coliforms	T: $b^*((L - P)/(J - P))^{-1/n'}$	$mg \cdot NH_2Cl \cdot L^{-1} \cdot min^{-1}$
C NH_2Cl for total coliforms	U: S/tc	$mg \cdot L^{-1}$
C NH_2Cl for fecal coliforms	V: U/tc	$mg \cdot L^{-1}$
Cl_2/NH_4 ratio	3.94	$mg \cdot mg^{-1}$
Cl_2/NH_4 ratio	0.34	$mole \cdot mole^{-1}$
Cl_2/NH_2Cl ratio	1.37	$mg \cdot mg^{-1}$
Chlorine decomposition rate	0.042	$mg \cdot min^{-1} \cdot L^{-1}$
Required total residual chlorine (combined chlorine)	X	$mg \cdot L^{-1}$
Monochloramine potential	PM: H*51.5/18	$mg \cdot L^{-1}$
Required HOCl for chloramine	HC: PM*52.5/51.5	$mg \cdot L^{-1}$
Cl_2/HOCl ratio	1.35	$mg \cdot mg^{-1}$
Cl_2 required to form HOCl	5.88*1.35 = 7.94	$mg \cdot L^{-1}$
Cl_2 injected to obtain the reduction of total coliforms	W: U/1.37	$mg \cdot L^{-1}$
Cl_2 injected to obtain the reduction of fecal coliforms	X: V/1.37	$mg \cdot L^{-1}$
Total Cl_2 to reduce total coliforms	Y: ((7.94 + 0.042*tc) + (0.45*F) + (1.12*G))/O	$mg \cdot L^{-1}$
Total Cl_2 to reduce fecal coliforms	Z : ((X + x) + (0.45*F) + (1.12*G))/O	$mg \cdot L^{-1}$
Equivalent in HOCl for total coliforms	Y*51.5/71	$mg \cdot L^{-1}$
Equivalent in HOCl for fecal coliforms	Z*51.5/71	$mg \cdot L^{-1}$
N-NH_3 eliminated	PM*14/51.5	$mg \cdot L^{-1}$
Residual N-NH_3	H – (PM*14/51.5)	$mg \cdot L^{-1}$

Table 8.22. *Disinfection with chloramines of wastewater*

8.5.3.5.2. Example of calculations for disinfection with chloramines

Table 8.23 provides an example calculation for the disinfection of wastewater using chloramines.

Parameters	**Calculations**	**Units**
Inlet average flow	86,667	$m^3 \cdot d^{-1}$
SS	5.0	$mg \cdot L^{-1}$
Temperature	20	°C
pH	7	–
BOD5	27	$mg \cdot L^{-1}$
Fe	0	$mg \cdot L^{-1}$
N-NO_2	0	$mg \cdot L^{-1}$
N-NH_3	5.0	$mg \cdot L^{-1}$
Total coliform inlet	7.20E+07	MPN/100 mL
Fecal coliform inlet	2.00E+06	MPN/100 mL
Required total coliform outlet	2.00E+02	MPN/100 mL
Required fecal coliform outlet	100	MPN/100 mL
Log reduction required for total coliforms	5.56	–
Log reduction required fecal coliforms	4.30	–
Contact time to obtain free chlorine	60	min
Homogenization coefficient of the contact tank	0.8	–
SS effects on microorganisms		
Fecal coliforms	5.9	MPN/100 mL
Total coliforms	21.1	MPN/100 mL
Enterococci	2.3	MPN/100 mL
Concentration in NH_3 and NH_4		
NH_3	6.07	$mg \cdot L^{-1}$
NH_4	6.43	$mg \cdot L^{-1}$
	0.36	$mmole \cdot L^{-1}$
Bacterial density coefficients in SS		
n total coliforms	2.8	–
b total coliforms	4	–
n′ fecal coliforms	2.8	–
b′ fecal coliforms	3	–
Calculation of concentrations in chlorine and chloramines		
Ct required for NH_2Cl for total coliforms	401.55	$mg \cdot min^{-1} \cdot L^{-1}$
Ct required for NH_2Cl for fecal coliforms	105.33	$mg \cdot min^{-1} \cdot L^{-1}$
C NH_2Cl for total coliforms	6.69	$mg \cdot L^{-1}$ NH_2Cl

Parameters	Calculations	Units
Calculation of concentrations in chlorine and chloramines		
C NH_2Cl for fecal coliforms	1.76	$mg \cdot L^{-1}$ NH_2Cl
Cl_2/NH_4 ratio	3.94	$mg \cdot mg^{-1}$
Cl_2/NH_4 ratio	0.34	$mole \cdot mole^{-1}$
Cl_2/NH_2Cl ratio	1.37	$mg \cdot mg^{-1}$
Chlorine decomposition rate	0.0283	$mg \cdot min^{-1} \cdot L^{-1}$
Residual total chlorine (combined chlorine)	1	$mg \cdot L^{-1}$
Available monochloramine	14.31	$mg \cdot L^{-1}$
Required HOCl for chloramine	14.58	$mg \cdot L^{-1}$
Cl_2/HOCl ratio	1.35	–
Cl_2 required to form HOCl	19.69	$mg \cdot L^{-1}$
Cl_2 injected to obtain the reduction of total coliforms	9.17	$mg \cdot L^{-1}$
Cl_2 injected to obtain the reduction of fecal coliforms	3.21	$mg \cdot L^{-1}$
Total Cl_2 total for reduction of total coliforms	26.73	$mg \cdot L^{-1}$
Total Cl_2 total for reduction of fecal coliforms	5.26	$mg \cdot L^{-1}$
Equivalent in HOCl for total coliforms	19.4	$mg \cdot L^{-1}$
Equivalent in HOCl for fecal coliforms	3.81	$mg \cdot L^{-1}$
N-NH_3 eliminated	4	$mg \cdot L^{-1}$
Residual N-NH_3	1	$mg \cdot L^{-1}$

Table 8.23. *Example of calculations of disinfection with chloramine in wastewater*

8.5.3.5.2. Conclusions on disinfection with chlorine and its derivatives

The microbiological safety of treated wastewater for irrigation is one of the most important factors to consider for this application, which is the most used in reuse. Chlorine has been used as a disinfectant since the early 19th century and is one of the most popular and widely accepted disinfectants for water and wastewater treatment due to its oxidative power, ease of use and low cost (0.0003–0.006 €/m^3).

However, chlorine is highly reactive with organic matter and causes the generation of organohalogen by-products (e.g. trihalomethanes and haloacetic acids) whose final concentrations may be higher than those required by local regulations.

The tolerance of water-borne pathogens to chlorine raises the issue of its choice for some applications where maximum disinfection is required for certain agents such as protozoa, some types of viruses and helminth eggs.

Eukaryotic pathogens are not satisfactorily eliminated with chlorine. *Cryptosporidium* and *Giardia lamblia* are reported to be highly chlorine tolerant. Some viruses have been shown to be resistant to chlorine and their contribution to waterborne diseases has sometimes been dramatic. For this reason, many developed countries are increasingly switching to alternative disinfectants such as UV irradiation. The bactericidal effectiveness of ClO_2 depends on several factors, such as dosage, contact time, water temperature, pH and organic matter. Unlike chlorine, chlorine dioxide does not hydrolyze in water.

Chlorine dioxide is a broad spectrum microbicide over a wide range of pH. It is more effective against viruses such as poliovirus and coliphages.

At pH 6.5, for a dose equivalent to chlorine, its germicidal effectiveness is much higher and with shorter contact times. At pH 8.5, chlorine dioxide maintains the same inactivation rates while chlorine will require higher doses. It is not reactive with ammonia and most nitrogen-containing compounds and is effective at lower doses than chlorine. It destroys phenolic compounds, simple cyanides and sulfides by oxidation. For odor control, chlorine dioxide will oxidize sulfides without formation of colloidal sulfur. It is also used to oxidize iron and manganese compounds.

Chlorine dioxide (ClO_2) is defined as a potential alternative to chlorine. The main advantage of ClO_2 over chlorine is its higher effectiveness and the formation of fewer by-products. However, its use may result in the presence of chlorites (ClO_2^-) and chlorates (ClO_3^-).

ClO_2 shows superior oxidation and bactericidal capacity than chlorine, but at a higher cost.

Chloramines are useful on partially nitrified effluents to avoid adding ammonia in the water to be disinfected. But residual NH_3 can be a limit of use for some applications (Dorn 1974).

8.6. Disinfection with ozone

Ozone is widely used for the disinfection of water intended for human consumption. It is indeed the best disinfectant among the oxidizing reagents available. Its efficiency gain compared to chlorinated compounds is especially remarkable on viruses and cysts of parasites such as *Giardia lamblia* and *Cryptosporidium*.

Ozone can be used at different points in the treatment process for different objectives. It is mainly used as an oxidant in pretreatment. It is used as a disinfectant and oxidant in inter-ozonation and post-ozonation.

8.6.1. *General information on ozone*

Ozone, chemical formula O_3, is a naturally occurring gas in the upper atmosphere, where it forms from oxygen under the action of ultraviolet radiation. Its presence, indicated by its characteristic odor, is sometimes detectable at lower altitudes in stormy weather (action of atmospheric electric discharges on the oxygen in the air).

Ozone is used in wastewater treatment for its energetic oxidizing properties. In aqueous solution, it is relatively unstable: under the pH conditions usually encountered in water treatment (pH 6–8), its stability is about 10–20 min.

This disappearance of ozone in water is explained by a natural phenomenon of radical self-decomposition initiated by OH^- ions, as well as by the consumption by oxidant consuming substances present in water (organic matter, oxidizable mineral matter).

The ozone demand of a water is the amount of ozone to be introduced to detect a non-zero residual after 5 min of contact time.

The instability of ozone in water results in a lack of residual power. Disinfection by ozonation must therefore, in most cases, be supplemented by the application of another reagent to relay the action of ozone. Usually, a small amount of chlorine or chlorine dioxide is sufficient to ensure bacteriostatic effect in the distribution network. There are, however, cases in which ozone is used alone: water of exceptional quality, short networks and/or in good condition, etc.

In addition to its exceptional disinfectant power (destruction of pathogenic germs 20–30 times faster than with chlorine), ozone is used in wastewater for its many properties such as:

– elimination of certain pesticides;

– complete elimination of phenols;

– high-proportion algae destruction;

– elimination of organic color.

Ozone has no noticeable effect on ammonia (very slow oxidation kinetics).

8.6.2. *Preparation and implementation of ozone*

In the field of water treatment, ozone is prepared by the action of an electric arc on a current of oxygen. The source of oxygen used is usually air or oxygen-enriched air. In some cases, pure oxygen will be used. The ozone thus produced is always diluted in the carrier gas used for its preparation: the concentrations of ozone obtained from air or oxygen are, respectively, 10–20 g per m^3 of ozonized air or 40–60 g per m^3 of ozonized oxygen.

An ozone installation consists of four parts:

– air treatment;

– ozone generation;

– ozone transfer to the water to be treated;

– recovery and treatment of ozonized vents.

8.6.2.1. *Air treatment*

The air used for ozone production must be perfectly dusted and extremely dry (dew point below –50°C), for production efficiency, as well as for the preservation of materials in the composition of ozonators. During the formation of ozone, the ionization of air forms nitrogen oxides, including nitric anhydride, which will transform into a highly corrosive acid in the presence of moisture. Furthermore, since the decomposition of ozone is accelerated at high temperatures, working with cold air is necessary to improve production efficiency.

Regardless of the technology chosen for air treatment, high-pressure channel (6–7 bar) or low-pressure (LP) channel (1.5–2 bar), the devices include:

– a step of overpressure or compression after filtration of ambient air;

– a desiccation step, which may include a first condensation step followed by a second drying step on an adsorbent.

The choice between high-pressure and LP process results from an optimization of investment and operating costs. In practice, high-pressure systems are implemented on low flow plants and low pressure systems on high flow plants.

8.6.2.2. *Generation of ozone*

Ozone is unstable and must be produced at the site of use. An industrial ozonator is a set of elementary generators, consisting of two electrodes:

– a stainless steel central electrode;

– a mass electrode made of a water-cooled stainless steel tube.

The space between these two electrodes is occupied by two dielectrics, one of which is very thin glass with high permittivity and the other air or oxygen. When these two electrodes are energized (for a voltage above a threshold value), the gas becomes ionized and electrically conductive: light discharges can be observed, which support the formation of ozone. Ozone production (expressed in kg of ozone/h) can be adjusted by adjusting the applied voltage.

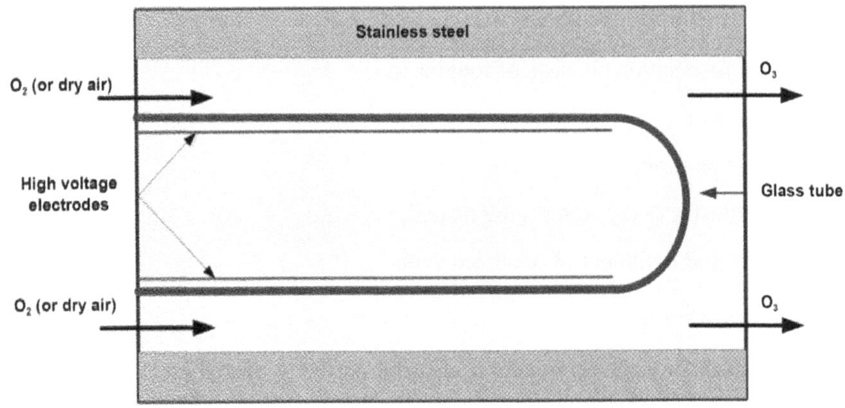

Figure 8.17. *Diagram of an elementary tube ozonator*

Most of the energy is dispersed as heat, which is discharged through the cooling water circulating around the ground electrode. Cooling water is generally water taken from the plant's service water circuit and put back in the treated water tank. If the content of this water is somewhat high in chlorides, it is necessary to provide a closed circuit of cooling water to prevent the stainless steel from being attacked.

8.6.3. *Transfer of ozone in water to be treated*

In general, the transfer of ozone into water is achieved by various processes: self-priming turbines, hydro ejectors or emulsifiers and diffusion of ozonated gas by means of porous diffusers. In the case of disinfection (post-ozonation), as the water is clarified and in particular cleared of SS, it is the technique using porous diffusers that is generally used. In the case of pre-ozonation, given the risk of clogging of diffusers, ozone is usually transferred by means of a turbine or an emulsifier.

Ozonation tanks are civil works of 4–6 m height and 10–15 m length on average, sized according to the maximum ozone rate to be applied. These tanks are separated into several compartments, where water and ozone flow co-current or countercurrent, which improves the diffusion of gas in the liquid.

The tank configuration is generally as follows (Figure 8.18):

– first tank operating countercurrent, designed for a contact time of 2 or 4 min: this is the oxidation compartment, receiving 60% of the overall level of injected ozone;

– second tank operating co-current or countercurrent (if there is an intermediate baffle), designed for a contact time of 4 or 6 min, into which 40% of the overall ozone level is injected;

– in some cases, both tanks into which ozone is injected are followed by a third contact tank.

If the installation is made of porous diffusers, ozonated air is injected into the first two compartments, usually at 70% of the airflow in the first one and 30% in the second one.

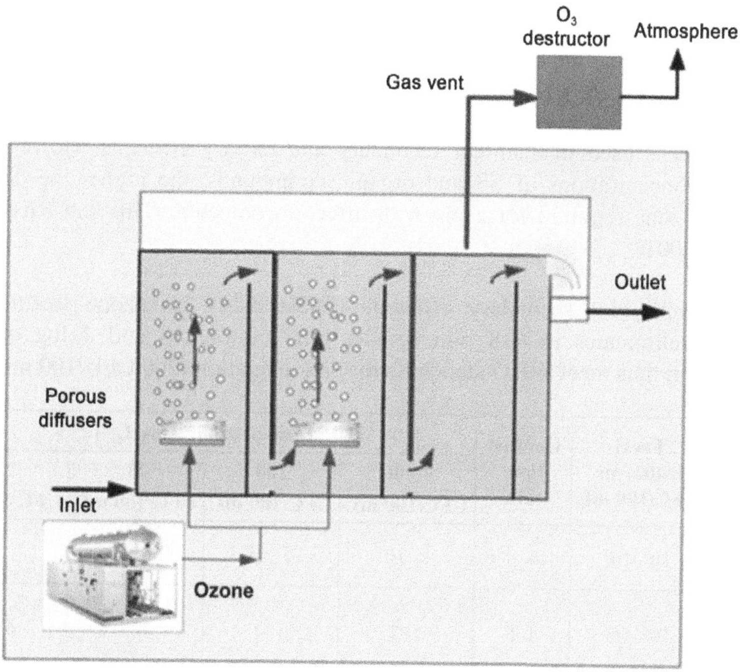

Figure 8.18. *Implementation of ozonation in reaction tanks*

Ozone diffusion by static mixer provides the best control of Ct, as:

– the entire water flow is brought into contact with ozone;

– ozone transfer to water is very fast, about a second;

– the fluid under consideration is monophasic, which enables a laminar flow in the structures.

Disinfection is carried out in concrete contact chambers of 4–6 m deep.

8.6.4. *Performance of disinfection with ozone*

The major interest of ozone lies in its very high potential for inactivation of viruses and bacteria, which constitute a major health problem in reuse processes. Ozone is more effective than chlorination and UV irradiation for protozoa inactivation.

Ozone is a powerful oxidant, and it is accepted that the bacteria are directly reduced by cell lysis. Ozone efficiency also depends on the quality of the wastewater to be disinfected. Organic substances involved in BOD, COD and TOC may have an immediate demand for O_3. The same is true for humic acids that affect the decomposition rate of the oxidant. As with other disinfectants, SS, iron, manganese, and nitrites increase the demand for ozone.

Ozonation is used to disinfect secondary and tertiary effluents. However, the higher the concentrations of SS and organic compounds, the higher the doses of transferred ozone required for a given disinfection objective (Hill and Rice 1982; Finch et al. 2001).

For example, for secondary effluents (Table 8.24), an ozone treatment of 5–10 mg/L eliminates in 6–8 min contact time between 3 and 4 log of fecal coliforms, and thus meet WHO standards for park irrigation (200 CFU/100 mL).

Type of effluent	Fecal coliforms UFC/100 mL	Contact time min	Dose of ozone ($mg \cdot L^{-1}$)			
			1,000 UFC/100 mL	200 UFC/100 mL	23 UFC/100 mL	2.2 UFC/100 mL
Activated sludge	10^5–10^6	6–8	2–10	5–10	7–12	–
Sand filtered effluent	10^4–10^6	4–5	1–2	3–4	4–5	4–6
Sand + GAC filtered effluent	10^3–10^5	4	1–1.5	1–2	2–3	3–4

Table 8.24. *Ozone dosage guide depending on the type of effluent and objectives*

Higher doses, from 4 to 6 mg·L^{-1}, ensure compliance with the Californian standard *Title 22* (2.2 Ct per 100 mL) on a sand filtered effluent. This amply justifies the interest of this filtration step before disinfection for particular uses.

Cryptosporidium cysts are more difficult to eliminate. For ozonation, Table 8.25 shows some of these values for systems measuring exclusively the residual ozone at tank outlet.

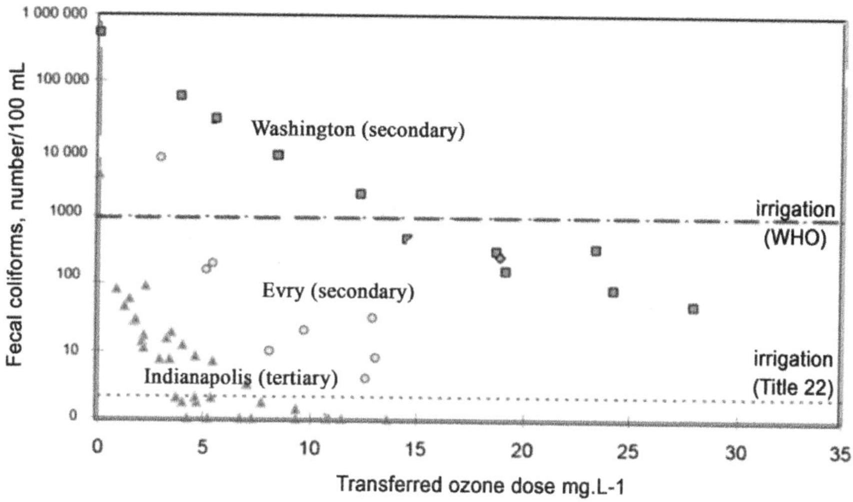

Figure 8.19. *Reduction of fecal coliforms with ozone on secondary effluents and on a filtered effluent*

In the case of ozone, C and t concentrations have equivalent weights (doubling the value of one and decreasing the other by a factor of 2 results in the same degree of inactivation). Ct generally implemented is 1.6 mg·min^{-1}·L^{-1} (or 0.4 mg·L^{-1} maintained for 4 min). For this value, with water at 20°C, the degree of inactivation is 3.0 log of *Giardia* cysts.

In the case of *Cryptosporidium*, the same Ct gives less than 0.5 log at 20°C.

The ozone concentration considered is the residual at the outlet of the ozonation tank. Ozonation is not effective at inactivating helminth eggs. Many parameters influence the performance of wastewater ozonation:

– Temperature: when the temperature rises, ozone's disinfectant power increases. Temperature influences the inactivation of *Cryptosporidium*, with a 4.5-fold increase in Ct for a 10°C rise in wastewater temperature.

– pH: changes in the pH of water alter the balance of available O_3 and OH. An increase in pH from 6 to 9 reduces the amount of O_3 available for disinfection.

– Suspended solids: the presence of SS may imply a higher consumption of O_3 and reduce the disinfection capacity of the oxidant if the dosages are not regulated.

– Other: natural organic matter (humic acids), BOD5, bromides, iron, manganese, nitrites, synthetic organic compounds and alkalinity all have ozone demand.

Reduction (log)	**Water temperature (°C)**			
Cryptosporidium	**1**	**5**	**10**	**20**
0.5	12.8	8.7	5.4	2.1
1	25.6	17.5	10.9	4.2
2	51.2	35.0	21.7	8.4

Table 8.25. *Ct (mg·L^{-1}·min^{-1}) for the inactivation of* Cryptosporidium *cysts with ozone*

These elements involve immediate demand for oxidants and therefore it is important to know the concentrations of these elements in order to determine the optimal O_3 doses for the disinfection step. In addition, another important parameter for disinfection is the optimization of the ozone transfer rate. It should be greater than 85% in all cases.

In summary:

– Ozone disinfection requires improved ozone transfer to water, correct dosage and knowledge of the quality parameters of the water to be treated.

– Ozone disinfection of wastewater is effective against a wide range of pathogens, including protozoa. High Ct values are required based on the initial pathogen concentration and final quality objectives. It is recommended that a physical barrier be provided upstream of this step to eliminate interference due to SS.

– A wide range of organic contaminants in wastewater can be degraded by ozone (see Chapter 2 of Volume 2 of this book).

– Ozone is widely used in wastewater treatment. Its ability to efficiently oxidize a wide range of contaminants and inactivate a wide range of microorganisms makes it a preferred reagent. On the other hand, its operating and maintenance costs are higher than those of other oxidants.

8.7. Organic peracids: performic acid and peracetic acid

Peroxides have the special feature of containing the peroxo group (-O-O-), which is in the case of organic peroxides linked to a carbon atom directly or through another atom. Organic peroxides can be considered as hydrogen peroxide (H_2O_2) derivatives, where organic groups replace one or more hydrogen atoms (Swern 1949).

Organic peracids contain the peroxycarboxylic group, COOOH, attached to an organic side chain that can be aliphatic, aromatic or cyclic.

Organic peracids have the highest oxidation potential of all peroxide organic compounds. Peracids are used as disinfectants, sterilizers, bleaching agents and as reagents in the production of certain chemical products.

8.7.1. *Production organic peracids*

A solution of peracid can be prepared by mixing a carboxylic acid and hydrogen peroxide according to reaction 1, which is one of the most widely used industrial synthesis routes (Klenk et al. 2005):

$$R\text{-}COOH + H_2O_2 \rightarrow R\text{-}COOOH + H_2O \quad (1)$$

The reaction may be catalyzed by an acid, for example, sulfuric acid, ascorbic acid or carboxylic acid esters.

However, performic acid (PFA) can be prepared without additional catalyst if an excess of formic acid (FA) is used in relation to the amount of hydrogen peroxide. The composition of the solution is affected by the initial molar ratios of the raw materials and the hydrogen peroxide concentration.

Aqueous peracids such as peracetic acid (PAA) can be stabilized to increase their shelf life. Commercial solutions are stabilized with compounds such as alkali metal polyphosphates, dipicolinic acid or quinoline derivatives (Gunther et al. 1969; Block 2001). In addition, the safety risks of peracids can be reduced with additives such as water, certain solvents (aliphatic and halogenated hydrocarbons) and salts (sulfates, phosphates, borates, silicates and carbonates).

Commercially available PFA (Desinfix, Kemira) consists of a reagent (35–50% H_2O_2 and 70–90% FA).

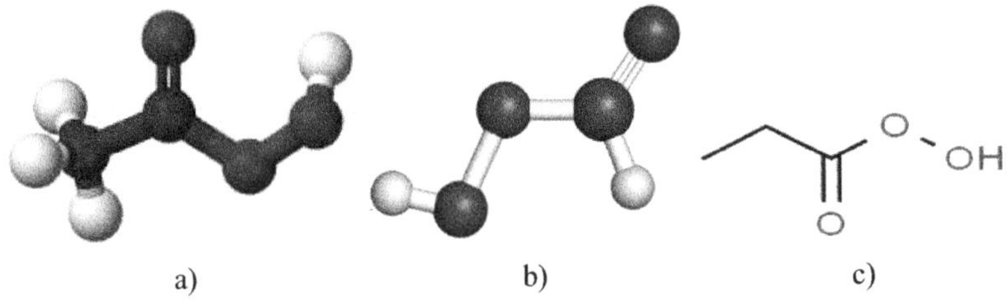

Figure 8.20. *(a) Peracetic acid (PAA); (b) performic acid (PFA); (c) perpropionic acid (PPA). For a color version of this figure, see www.iste.co.uk/gaid/reuse1.zip*

Another frequently used method of synthesis is to create a reaction between carboxylic anhydride and hydrogen peroxide (reaction 2), which produces an anhydrous solution of peracid in the carboxylic acid (Mattila and Aksela 2000; Yousefzadeh et al. 2014).

This reaction can be catalyzed by inorganic bases or metal salts:

$$R_1\text{-COOOC-}R_2 + H_2O_2 \rightarrow R_1\text{-COOH} + R_2\text{-COOOH} \quad (2)$$

8.7.2. *Disinfection and oxidation mechanisms of organic peracids*

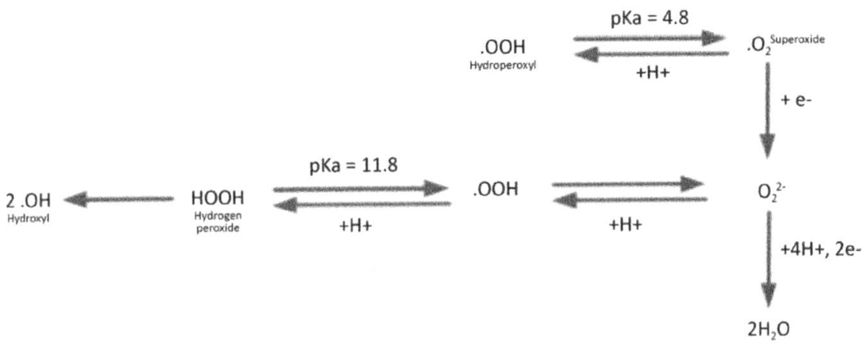

Figure 8.21. *Reaction mechanisms of peracids*

The mechanisms of disinfection and oxidation of PAA are based on the formation of highly oxidizing radical species such as hydroxyl radical (HO·), superoxide anion ($\cdot O_2^-$) and hydroperoxyl radical ($HO_2\cdot$). These radicals are called reactive oxygen species, or active oxygen (Figure 8.21).

The formation of radicals begins with the oxygen/oxygen bond rupture (reaction 3), which is the stage determining the speed in the chain reaction (Luukkonen et al. 2014, 2015). The reaction requires activation to occur, such as a small amount of a catalyst made up of a transition metal or UV irradiation (Bianchini et al. 2002a, 2002b).

Subsequently, in the case of PAA, the following aqueous chain reactions take place (reactions 4–9):

$$CH_3COOOH \rightarrow CH_3COO\cdot + HO \quad (3)$$

$$CH_3COOOH + HO\cdot \rightarrow CH_3CO\cdot + O_2 + H_2O \quad (4)$$

$$CH_3COOOH + HO\cdot \rightarrow CH_3COOO\cdot + H_2O \quad (5)$$

$$CH_3COO\cdot \rightarrow H_3C\cdot + CO_2 \quad (6)$$

$$2CH_3COO\cdot \rightarrow 2H_3C\cdot + 2CO + O_2 \quad (7)$$

$$H_3C\cdot + O_2 \rightarrow CH_3OO \quad (8)$$

$$CH_3COO\cdot + HO\cdot \rightarrow CH_3COOOH \quad (9)$$

All generated radicals contribute to oxidation, and $HO\cdot$ and H_3C play the most important roles in the mechanism (Bach et al. 1996; Rokhina et al. 2010). In PAA catalytic oxidation, $\cdot CH_3$ is scarce due to the lower reaction rate constant. Furthermore, organic radicals have longer half-lives than $HO\cdot$ and are effective in their antimicrobial action (Flores et al. 2014).

The most common damage to biomolecules caused by PAA is:

– oxidation of sulfhydryl (-SH) and disulfide (S-S) elements;

– oxidation of double bonds in proteins, enzymes and other biomolecules;

– inactivation of the catalase enzyme, which inhibits the hydroxyl radical by oxidation;

– disruption of the chemiosmotic function of the cytoplasmic membrane;

– denaturation of proteins;

– reaction on DNA molecules;

– oxidation of enzymes. Biochemical pathways, transport through membranes and intracellular solute levels are impaired.

In the case of viruses, inactivation occurs through damage to surface structures such as protein envelopes or attachment to sites required for infection of host cells (Koivunen and Heinonen-Tanski 2005a). The damage caused by PAA is irreversible, meaning that microbes are unable to repair damaged biomolecules, and there is no revival (Antonelli et al. 2006). Hydrogen peroxide serves as an additional source of hydroxyl radicals.

8.7.3. *Performic acid*

PFA, developed by Kemira Oyj (Desinfix process), is a solution prepared in situ by mixing FA and hydrogen peroxide (H_2O_2) at a constant ratio. Due to its instability, PFA is produced on site shortly before use.

The PFA solution is produced by the following exothermic reaction:

– main reaction:

$$H_2O_2 + HCOOH \rightarrow CO_2 + H_2O$$

– intermediate steps:

$$HCOOH + H_2O_2 \rightarrow HCOOOH + H_2O$$

$$H_2O_2 \rightarrow H_2O + \frac{1}{2}O_2$$

– decomposition steps:

$$HCOOOH \rightarrow HCOO\cdot + HO\cdot \rightarrow CO_2 + H_2O$$

$$HCOOOH \rightarrow HCOOH + \frac{1}{2}O_2$$

PFA is prepared in proportions of FA (70–80% by weight) and H_2O_2 (35–50% by weight) in the presence of sulfuric acid as a catalyst.

PFA is a broad-spectrum disinfectant that can inactivate viruses, bacteria and bacterial spores, mycobacteria, as well as microscopic fungi.

As PFA is an unstable compound, it is prepared in situ upstream of the disinfection reactor. The reaction steps produce radicals that are part of the disinfection performance. The complete reaction produces carbon dioxide and water.

8.7.3.1. *Disinfection with performic acid*

PFA is a broad-spectrum disinfectant that can inactivate viruses, bacteria and bacterial spores, mycobacteria, as well as microscopic fungi.

PFA solutions for disinfection are prepared by mixing 70–90% by weight of FA and 35–50% by weight of hydrogen peroxide (H_2O_2) in the presence of a catalyst. Due to its instability, PFA cannot be stored and is therefore produced in situ just before use.

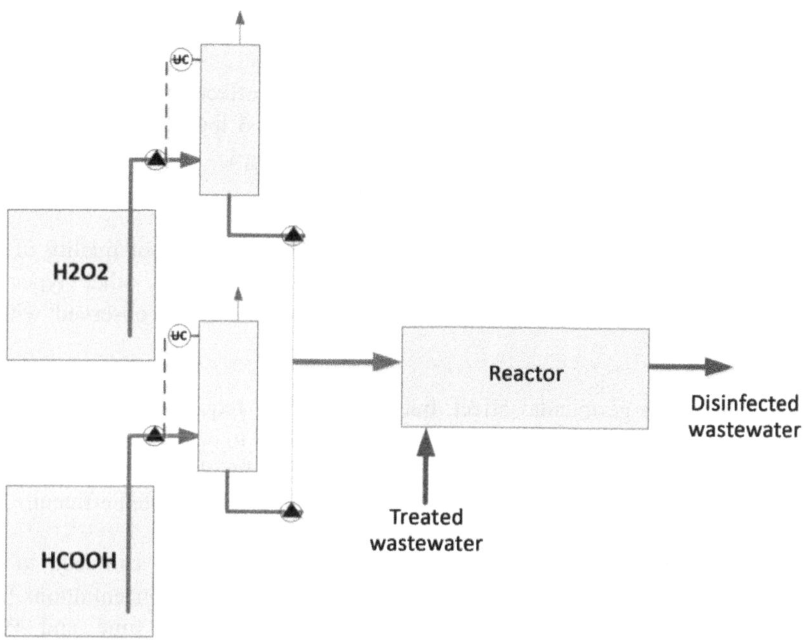

Figure 8.22. *Disinfection of wastewater with performic acid*

Desinfix (DEX 135, new generation at Kemira) was developed to improve efficiency, reliability and implementation. The mixing unit is a thermostatic bath tubular reactor and the resulting PFA concentration is between 12 and 15% by weight (DEX 135). The system is equipped with UPS and leak alarm; all variables such as temperature, pressures, levels and flows are controlled, and the system provides for production shutdown and emergency washing of appliances for all needs.

Operation and maintenance do not require particularly specialized skills, but technical training is advised (Mattila and Aksela 2000; Gehr et al. 2009).

8.7.4. *Peracetic acid*

PAA is a germicidal agent, particularly used in the food industry and as a disinfectant in hospitals. For wastewater treatment, its use is recent. It is sold in different solution mixtures with H_2O_2 and acetic acid (solutions at 15–20%):

$$CH_3CO_2H + H_2O_2 \rightarrow CH_3CO_3H + H_2O$$

PAA is a powerful disinfectant for bacteria, but less effective against viruses. A maximum reduction of 3.8 log of *Escherichia coli* and 4 log of total coliforms is achieved on a secondary effluent with a dosage of 15 $mg \cdot L^{-1}$ and a contact time of 36 min (Mezzanotte et al. 2007).

Its main advantage is that it remains effective despite the poor quality of the wastewater to be treated, unlike what is observed with all other types of disinfectants. In particular, there are no significant differences observed with a variation of SS.

PAA has a high germicidal effect, but high efficacy requires a contact time of about 30 min. PAA produces little by-products compared to ozone and chlorine. The PAA doses required are higher than those of chlorine for the same contact time. Note that the use of PAA increases the level of COD in the disinfected effluent.

In the presence of chlorides and organic substances, PAA can lead to the formation of organochlorine compounds, even if they are in low concentrations. The parameters influencing disinfection performance are contact time and PAA concentration.

8.7.5. *Degradation kinetics of peracids*

The degradation kinetics of organic peracids takes into account dwell time, dose of peracid, temperature, pH, amount of organic matter, presence of transition metals and hardness (Sanchez-Ruiz et al. 1995; Lazarova et al. 1998; Howarth 2003; Falsanisi et al. 2006). In addition to the chemical reactions proposed above, three other pathways are involved in the consumption of peracids in aqueous media: immediate decomposition, hydrolysis and transition metal-catalyzed decomposition as shown in reactions 10–12.

At pH 5.5–8.2, the main reaction is immediate decomposition (reaction 10):

R-COOOH + H_2O → R-COOH + H_2O_2 (10)

2 R-COOOH → 2 R-COOH + O_2 (11)

R-COOOH → R-COOH + ½O_2 + other products (12)

The degradation rate of the peracid can be presented with the following differential:

$$\frac{dC}{dt} = -k\alpha\ C^{\alpha}$$

where:

– $k\alpha$ is the rate constant, α is the reaction order (0, 1 or 2);

– C ($mg\cdot L^{-1}$) is the peracid concentration.

Initial consumption of peracid (D, $mg\cdot L^{-1}$) occurring immediately after application can be taken into account by replacing the initial concentration of peracid (Co, $mg\cdot L^{-1}$) with the term Co – D, which translates into the following laws of integrated rates:

$$Ct = (Co - D) - ko\ t \qquad t\,1/2\ = \frac{1}{k2Co}\ (13)$$

$$Ct = (Co - D)e^{-k1t} \qquad t\,1/2\ = \frac{Ln21}{k1}\ (14)$$

$$Ct\ = \frac{Co-D}{1+k2t\ (Co-D)} \qquad t\,1/2\ = \frac{1}{k2Co}\ (15)$$

where *ko* ($mg\cdot L^{-1}\cdot min^{-1}$), k1 (1/min) and k2 ($L\cdot mg^{-1}\cdot min^{-1}$) are the rate constants of zero-order, first-order and second-order equations, respectively.

These models are widely used and describe the kinetics of microorganism inactivation. The models for disinfectant demands take into account the decrease in the disinfectant concentration during the contact time by introducing decomposition kinetic constants. These models derive from differential rate laws (Gyürék and Finch 1998):

$$\frac{dNt}{dt} = -k\beta\ mNt^{x}\ C^{n}t^{m-1}\ (16)$$

where:

– dNt/dt is the inactivation rate;

– Nt is the number of surviving microbes at contact time t (min);

– C is the concentration ($mg \cdot L^{-1}$);

– m, n and x are empirical constants (dimensionless);

– kβ is the disinfection rate constant.

The Chick–Watson model (Chick 1908; Watson 1908) is one of the oldest models of disinfection kinetics. It follows the first kinetic order, and the C × t values obtained from the model are frequently used in the design of disinfection systems:

$$Ln \frac{Nt}{No} = -\gamma \, C^n t \quad (17)$$

where:

– Nt is the quantity of microbes at t time (min);

– No is the initial quantity of microbes;

– γ is the rate constant.

Parameter n indicates n > 1: the dose is greater than the contact time; n = 1, the dose and contact time are of equal importance; n < 1, the contact time is greater than the dose.

8.7.6. *Primary, secondary and tertiary wastewater applications*

For an application on treated wastewater after secondary treatment (biological-sedimentation basin), the doses of PAA and the contact times required are 10–20 $mg \cdot L^{-1}$ and 30 min to achieve a 2–3 log reduction in the amount of enteric bacteria detected (Sanchez-Ruiz et al. 1995; Koivunen and Heinonen-Tanski 2005b).

With PFA, on the same type of wastewater, a dose of 3.4 $mg \cdot L^{-1}$ and a contact time of 45 min made it possible to obtain 3 log elimination of fecal coliforms.

For an application on treated wastewater after tertiary treatment (media filtration-disinfection or disinfection alone), PFA doses are 0.6–2 $mg \cdot L^{-1}$ for a contact time of 10 min (Karpova et al. 2013; Ragazzo et al. 2013).

With PAA, dosages and contact times were between 1.5 and 10 mg·L^{-1} and 10–36 min. Field results showed PFA was more effective than PAA, with lower dosages and shorter contact times (Ragazzo et al. 2013; Chhetri et al. 2014).

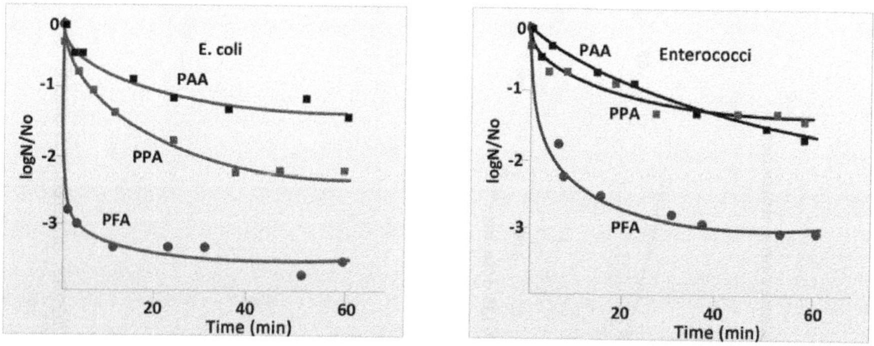

Figure 8.23. *Inactivation of* E. coli *and enterococci with PFA, PAA and PPA (dose 1.5 mg·L^{-1})*

Figure 8.23 illustrates the efficiency of three peracids at initial *E. coli* and enterococci concentrations of 2.90 × 10^4 and 1.84 × 10^3 CF per 100 mL, respectively.

The efficiency of peracids with *E. coli* is classified as PFA > PPA > PAA, while for enterococci it is classified as PFA > PAA ≈ PPA (Ragazzo et al. 2013; Chhetri et al. 2014).

Additional results with PFA (DEX-135, Kemira) on a secondary effluent are shown in Figure 8.24. The graphs are based on approximately 150 data points from eight different large-scale facilities, with SS concentrations below 20 mg·L^{-1} for each and between 30 and 40 mg·L^{-1} for others.

Bacterial concentrations were:

– total coliforms: 5 × 10^4 to 2 × 10^6 FC for 100 mL;

– *E. coli*: 5 × 10^4 to 2 × 10^6 FC for 100 mL;

– enterococci: 2 × 10^3 to 7 × 10^4 FC for 100 mL.

The effectiveness of DEX-135 depends on parameters such as water temperature, SS concentration, organic and mineral substances that can be oxidized, contact times, and applied doses.

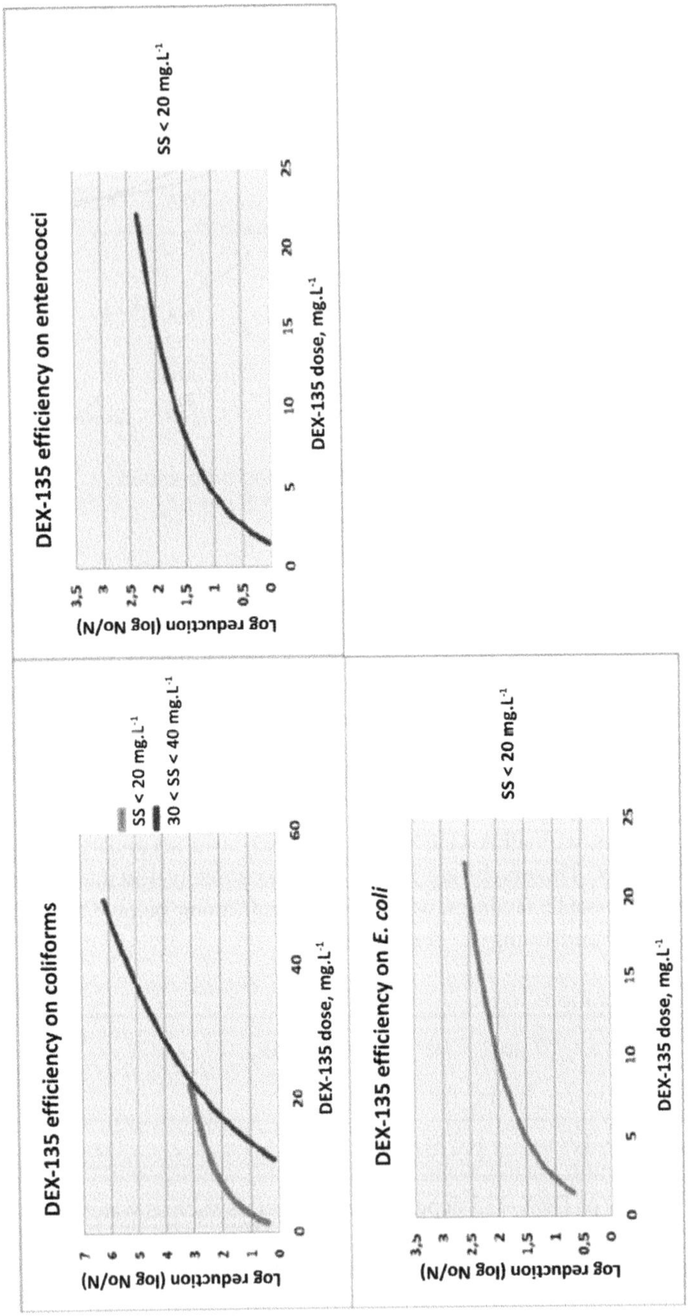

Figure 8.24. *Efficiency of performic acid (DEX-135, Kemira) to reduce coliforms, enterococci and* E. coli

Figure 8.24 shows that for wastewater containing an SS concentration of 20 mg·L^{-1}, a 2.5 log reduction on coliforms, 1.5 log on enterococci and 2 log on *E. coli* is obtained with a dose of 10 mg·L^{-1} and a contact time of 10 min.

The same results show that with an SS concentration between 30 and 40 mg·L^{-1}, a 2.5 log reduction is only obtained on coliforms with a dose at least equal to 20 mg·L^{-1}.

8.8. UV disinfection

8.8.1. *General information on UV radiation*

In the spectrum of electromagnetic waves, UV radiation is between X-rays and violet of the visible spectrum, in the 100–400 nm wavelength range (Figure 8.25). The UV spectrum is divided into four bands:

– UVA: 400–315 nm: skin pigmentation;

– UVB: 315–280 nm: production of antirachitic vitamin D;

– UVC: 280–200 nm: this region of the UV spectrum is used for its microbicidal power, with a maximum germicidal action at d = 253.7 nm (265 nm according to some authors);

– UV < 200 nm: radiation in this part of the UV spectrum induces ozone production from oxygen and is absorbed by most substances including air.

For disinfection to occur, the lamps must provide light radiation with a spectrum of emission in the UVC region, which is most effective at producing a germicidal effect.

Treatment involves exposing the wastewater to be treated to a source of UV radiation. UV radiation is electromagnetic waves with wavelengths between 100 and 400 nm. Their germicidal power depends on the wavelength emitted. UVC radiation (wavelength between 200 and 280 nm) is the most germicidal and should be taken into account in water disinfection. The optimal germicidal region is at 262 nm. Exposure of bacteria and viruses to UV radiation at near 262 nm wavelength results in deterioration of their cellular material (DNA and RNA). The evolution of knowledge means that importance is now also given to other wavelengths, which can induce more general cellular transformations. There is a separation of the double bonds of DNA by absorption of UV photons and creation of dimers that prevent replication.

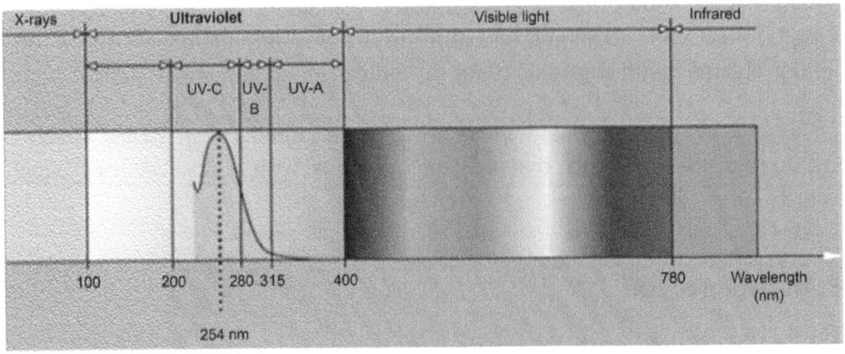

Figure 8.25. *Spectrum of ultraviolet and visible light. For a color version of this figure, see www.iste.co.uk/gaid/reuse1.zip*

8.8.2. *Inactivation mechanisms*

Two effects of UV are currently recognized: bactericidal and bacteriostatic. The bactericidal effect refers to a direct rupture of the cell membrane due to UV irradiation. The bacteriostatic effect concerns the modification of the genetic capital of irradiated microorganisms. It has been shown that UV radiation creates thymine dimers in the DNA chain, which result in dissociation of unsaturated double bonds. The cell can no longer reproduce or create the proteins necessary for its survival. It is therefore inactivated and no longer has an infectivity potential. Microorganisms have a UV sensitivity that is highly dependent on the absorption curve of their genetic material. Nevertheless, all the results obtained so far confirm the similarity between the DNA absorption curve (Shin et al. 2005) and spectral sensitivity for wavelengths between 230 and 290 nm. Other molecules can also absorb radiation (some proteins or enzymes, etc.) provided that this radiation is available for disinfection, because the majority of these molecules absorb in the low wavelength range, which are themselves heavily absorbed by water.

Compared to chemical disinfection processes, the technique using ultraviolet radiation, whose bactericidal properties have been known for many years, has the advantages of not inducing toxicity and not generating sapid by-products (like chlorine). However, it has no residual action.

The effectiveness of disinfection is also linked to the applied UV radiation dose, which is defined as the product of the UV intensity by the time germs are exposed to radiation.

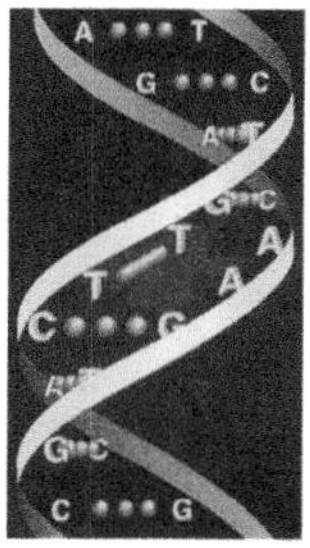

Figure 8.26. *Deterioration of cellular material with UV. For a color version of this figure, see www.iste.co.uk/gaid/reuse1.zip*

The usual configuration of the UV reactor is a low pressure mercury lamp with a quartz sleeve in direct contact with water; water enters the unit and flows along the annular space between the quartz sleeve and the chamber wall. Other configurations include lamps separated from water, for example, lamps surrounded by "bundles" of PTFE tubes through which the water flows.

Disinfection will only be efficient if a sufficient dose of UV is applied. The UV radiation dose is expressed in energy flux, units of $mWs{\cdot}cm^{-2}$, which is the product of the intensity emitted by the lamp and the dwell time of the water in the reactor. The minimum dose required for disinfection depends on several factors, including the sensitivity of microorganisms, but is generally considered to be between 16 and 40 $mWs{\cdot}cm^{-2}$.

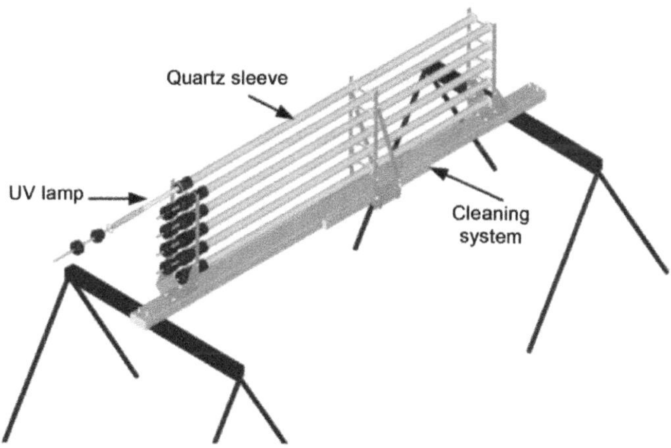

Figure 8.27. *UV lamp system. For a color version of this figure, see www.iste.co.uk/gaid/reuse1.zip*

8.8.3. *Lethal dose and inactivation kinetics*

The radiation absorption follows the Lambert–Beer law:

$$I = (P/S)\,(T/100)^{-d}$$

with:

– P: UV power emitted by the lamp (W);

– S: surface at the distance d of the lamp = 2π(d-r)L, lamp ray r, L its length;

– d: water thickness (cm);

– I: intensity received at the distance d of the lamp.

The radiation emitted by UV lamps is partly absorbed by water so that microorganisms receive only part of the radiation.

The percentage of transmission over 10 mm of effluent (T%) is the predominant parameter for design. It is easily measured by means of a 254 nm spectrometer and a 1-cm thick quartz cell.

Thus, considering only the absorption term $(T/100)^{-d}$ (neglecting P/S dispersion), for a 3-cm water sheet, the UV dose emitted on the surface of the lamp is divided by 3 for a 65% UV transmission and by 15 for a 40% UV transmission!

The water thickness around the UV lamp must be thin to obtain the best UV reduction performance.

UV transmission has a very large influence on the size.

The lethal dose of exposure is calculated as:

$$D = \frac{Pt}{S} e^{-kx}$$

with:

– P: germicidal power of the UV source (in W);

– S: UV emitting surface (in cm^2);

– T: exposure time of a volume element (in s);

– k: UV absorption coefficient of the water to be treated (cm^{-1}): this coefficient varies from 2 to 10 m^{-1} (0.02–0.1 cm^{-1}) for drinking water;

– x: water thickness (cm);

– D: inactivation dose ($mJ \cdot cm^{-2}$).

The amount of energy received by a microorganism under UV radiation is equal to:

$$D = I \times t$$

with:

– I: power received per unit area, depending on the lamp power used, its geometry, the path covered by the UV radiation and the UV absorption coefficient in the water to be treated ($mW \cdot cm^{-2}$);

– t: exposure time in seconds;

– D: energy received per unit area or exposure dose expressed in $mWs \cdot cm^{-2}$ (or in $mJ \cdot cm^{-2}$).

Thus, the UV dose ($mJ \cdot cm^{-2}$) is the total energy of radiation from all germicidal wavelengths passing in all directions through an infinitesimal sphere of cross-sectional area dA (cm^2), divided by dA.

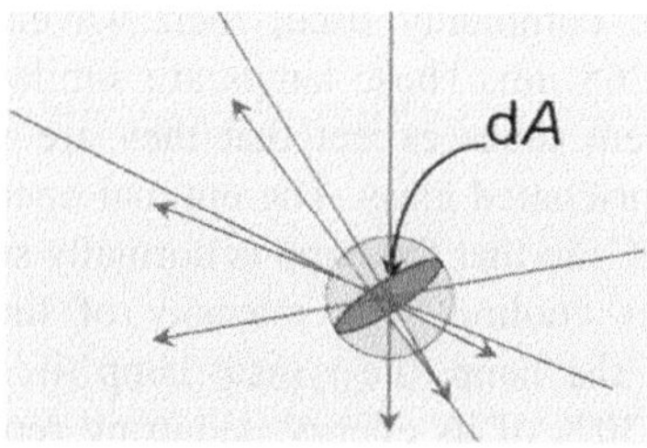

Figure 8.28. *Radiation on a spherical particle*

The microorganism inactivation law is:

$$\text{Log}\frac{\text{Initial concentration } No}{\text{Final concentration } N} = kD$$

The reduction in log units is therefore directly proportional to the dose. The inactivation coefficient k specific to each type of microorganism reflects the specific sensitivity of different germs to UV.

In fact, if turbulences are good, the probability that germs pass at the maximum lamp distance (e.g. by going along the edges) is low and the minimum intensity is replaced by the average intensity calculated by integrating into the reactor volume the intensity received at each point and emitted by the different lamps. Sizing is usually done at the lowest UV transmission and peak flow rate.

8.8.4. *Implementation*

The light sources used in UV disinfection are mainly based on mercury vapor lamps. The characteristic of mercury is that it has a resonance line at 253.7 nm, which is in the maximum disinfection efficiency band. In this type of lamp, the electric discharge between the electrodes causes the excitation of the mercury atoms, which, when they fall back into their ground state, emit radiation. The relative intensities of the different lines emitted depend on the mercury pressure in the lamp.

UV rays are produced by LP mercury lamps or medium-pressure mercury lamps that have a lower germicidal efficiency than LP lamps (BP).

8.8.4.1. *Low-pressure lamps*

Low-pressure mercury lamps, which generate 85% of their energy at 254 nm wavelength, are the most commonly used; their wavelength is in the optimal germicidal range of 250–265 nm. These lamps are similar in design, construction and operation to fluorescent tubes except that they are made of UV-transparent quartz rather than phosphor-coated glass. The optimal operating temperature of the lamp is approximately 40°C, so that the lamp is normally separated from water by a "sleeve" to prevent water cooling. The intensity of the emitted UV radiation decreases with the age of the lamp; the typical lamp life is about 14,000 h, after which it loses more than 30% of its energy, requiring replacement (Bukhari et al. 1999; Bolton and Karl 2003).

Given the limited power of these lamps, a large number must be installed to ensure proper disinfection (approximately 500 lamps per 1,000 m^3/h). The lamps are arranged in skids (modules including a given number of lamps) in open channels where water circulates.

The usual UV reactor configuration is an LP mercury lamp with a quartz sleeve in direct contact with water; water enters the unit and flows along the annular space between the quartz sleeve and the chamber wall.

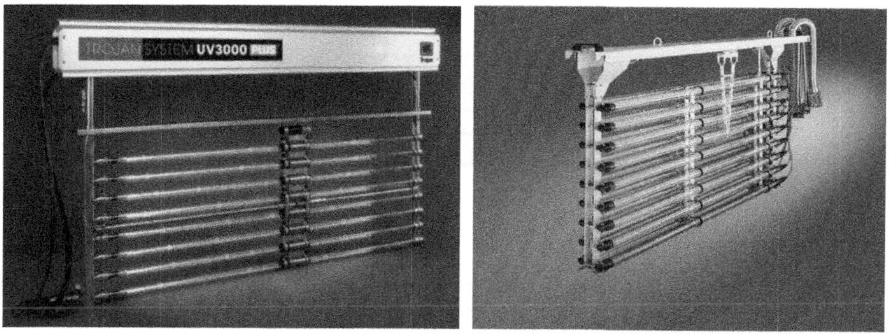

Figure 8.29. *UV lamps from the Trojan and Wedeco suppliers. For a color version of this figure, see www.iste.co.uk/gaid/reuse1.zip*

8.8.4.2. *"Medium and high-pressure" lamps*

These lamps have a wider emission spectrum, so worse UV efficiency (approximately 6–10%); however, it depends on whether only the specific radiation at 254 nm is considered or if other radiations around this value are also integrated, which have a lesser but not negligible efficiency (Gaid et al. 2005; Koivunen and Heinonen-Tanski 2005a).

These lamps have much higher unit powers than high-pressure lamps, above 2,000 W, so the number of lamps to be implemented is much lower (a few lamps).

The main advantages of low pressure are its good efficiency and monochromatic spectrum making it possible to target the germicidal wavelength at 254 nm and avoid the formation of by-products at UVC disinfection doses. The advantage of medium pressure is its power, which helps to process large flows with a minimum of lamps. However, medium-pressure lamps have the disadvantage of promoting the formation of unwanted by-products because of their polychromatic spectrum. In wastewater application, UV reactors are open or closed reactors.

Disinfection is carried out in a treatment chamber where the water is irradiated by one or more UV generators, as each generator is isolated from contact with water by a quartz jacket permeable to UV rays.

Disinfection is achieved by circulating a thin annular water sheet around these sleeves. The hydraulic regime of the flow in the chamber must be turbulent and piston-type to ensure good homogeneity of treatment (same exposure time for all germs).

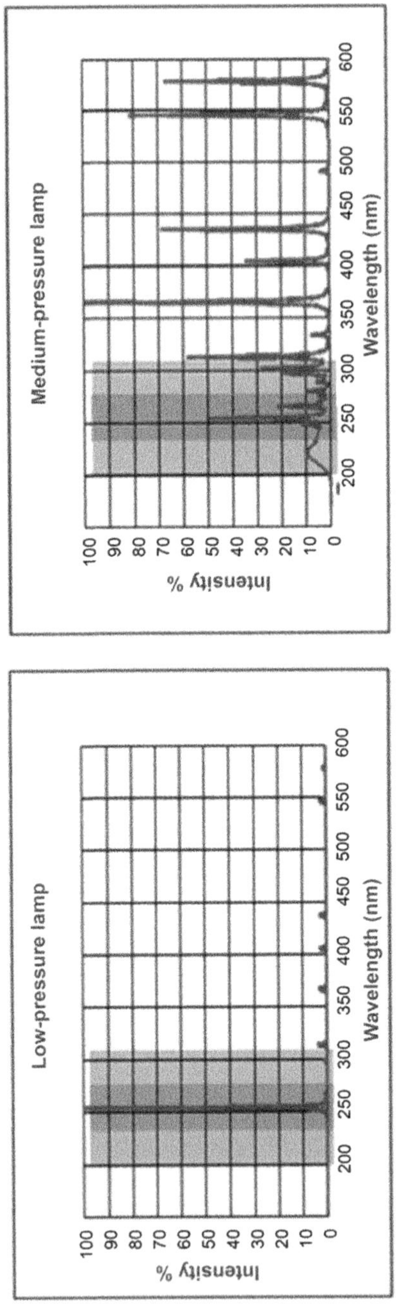

Figure 8.30. *Wavelengths of low- and medium-pressure lamps*

The installations are equipped with a radiometer in order to be able to monitor UV emission efficiency and detect any malfunctions.

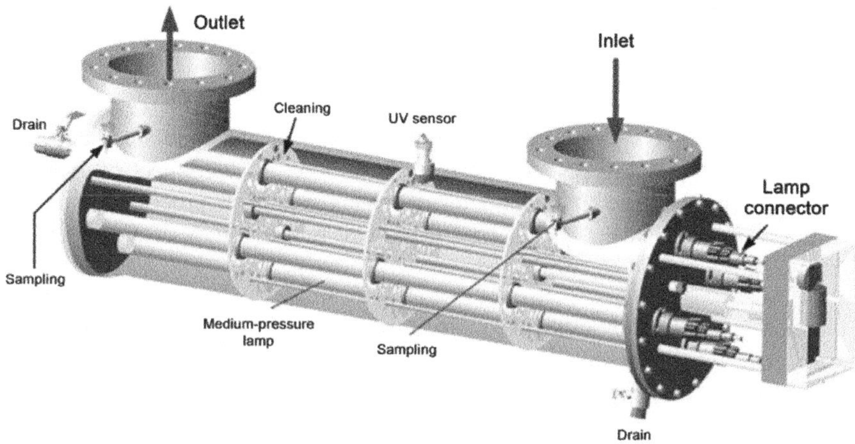

Figure 8.31. *Closed reactor used for drinking water. For a color version of this figure, see www.iste.co.uk/gaid/reuse1.zip*

The minimum dose required for disinfection depends on several factors, including the sensitivity of microorganisms, and is applied at 40 $mJ \cdot cm^{-2}$. Depending on the performance required and the type of microorganisms, this dose can vary between 40 and 70 $mJ \cdot cm^{-2}$. (1 $mWs \cdot cm^{-2}$ = 1 $mJ \cdot cm^{-2}$).

The UV reactor is mounted in line with the filtered water pipe. Cleaning can be performed automatically with the equipment already installed on the reactor or manually with a cleaning kit in place, made of a cleaning reagent container and a recirculation pump installed on a mobile trolley.

During cleaning, the reactor inlet/outlet valves are closed while the cleaning kit is connected to the cleaning nozzles. The cleaning product is recirculated for approximately 20 min. This is done approximately every 6 months.

The reactor can be bypassed in case of a necessary intervention. The chlorine then ensures the disinfection objectives.

Several new treatment technologies have been developed for the inactivation of *Cryptosporidium*. These include UV or white pulsed light systems and combined filtration-irradiation or adsorption-irradiation techniques, which increase exposure to UV.

UV pulse and white pulsed light devices generate high-intensity and short-duration radiation pulses, so they would have more effective inactivation of oocysts than conventional UV systems.

8.8.5. *Sizing parameters*

The sizing of a UV device gives rise in the first place to a measurement of the transmissivity by the UV resource at 254 nm through a water sheet of 1 cm, for example, or its estimate in relation to equivalent treatment processes and water qualities produced. For the reactor size, the lowest transmissivity of the water to be treated with UV should be considered.

The second parameter to be taken into account is the flow rate to be treated in the reactor. The maximum flow rate which can potentially pass through the reactor should be considered.

8.8.5.1. *UV dose (mJ·cm^{-2} or J·m^{-2})*

As mentioned above, it is the product between the intensity (W·m^{-2}) and the irradiation time. It is expressed either in mJ·cm^{-2} or in J·m^{-2}, 1 mJ·cm^{-2} = 10 J·m^{-2}.

The intensity depends on the type of lamp, its power, its age, the quality of the quartz sleeve and the quality of the water (transmittance).

8.8.5.2. *Electric power consumed*

8.8.5.2.1. Calculation of the electric power consumed in Wh/m^3

The relationship between UV dose and the electric power to be applied per m^3 of water to be treated (PUV) is determined by the following formula:

$$\text{PUV (Wh/m}^3) = \frac{\text{UV dose } K_{\text{iron}}}{R\ K_{\text{lamp}}}$$

K_{iron} is a coefficient related to the dissolved iron content of the water to be treated. For an iron concentration less than 0.5 mg·L^{-1}, $K_{iron} = 1$.

K_{lamp} is the remaining electrical power rate at the lamp end of life (after 12,000 h), equal to 85% (15% loss of power).

Kt is a ratio to transform J/m^2 into Wh/m^3:

$$Kt\ (\text{Wh/m}^3) = Y\ EXP\ (2{,}728\ T)$$

T is the transmittance (in %) and Y is a coefficient varying between 4 and 6, depending on the type of UV reactor.

8.8.5.2.2. Calculation of the number of lamps

The total number of lamps is equal to the total electrical power to be applied divided by the unit power of a lamp.

8.8.5.2.3. Calculation of the power installed (kWh)

$$\text{PUV (kW)} = Kt.\text{Number of lamps}/1{,}000$$

8.8.5.2.4. Calculation of the power consumed by m³ of water (kW·m⁻³·j⁻¹)

$$\text{PUV (kW.m}^{-3}.\text{j}^{-1}) = \text{PUVinstalled} \times 1{,}000/\text{Flow}$$

Water flow is expressed in m^3 per day.

8.8.6. *Factors influencing UV treatment efficiency*

The main factors affecting the design of a UV system for disinfection of water for reuse are UV transmission, SS, flow and/or hydraulics, iron, hardness, lamp aging, UV dose and wastewater pretreatment. All these factors affect the UV dose required to disinfect wastewater for reuse. The UV dose produced by a UV system is what disinfects the water. A standard method has been developed to determine this UV dose so that different systems can be compared and to ensure that the appropriate UV dose was applied to wastewater.

8.8.6.1. *Quality of the wastewater to be treated*

8.8.6.1.1. Wastewater transmittance

The quality of the wastewater subject to UV radiation is fundamental. The UV transmission of wastewater affects the intensity of the UV light produced by the UV system. The UV system should be designed for the minimum acceptable UV transmission expected in the treatment plant. Therefore, the UV treatment facility is sized according to the UV transmittance of the wastewater to be treated; the lower the UV transmittance, the larger the size/power of the UV treatment facility, and the higher its costs (installation, operation). The transmittance is inversely proportional to turbidity, SS, color (humic acids), residual iron and manganese dissolved.

The presence of these compounds can create a "UV demand" by analogy with a chlorine demand. The method used to quantify this interference is the transmittance measurement, which represents the amount of light available to irradiate a microorganism at a given distance.

8.8.6.1.2. Turbidity and suspended solids

Turbidity has the effect, in addition to decrease the penetration of UV rays into water, of protecting microorganisms from the action of these rays.

The SS act at two levels:

– they absorb the radiation directly and thus reduce the available UV dose;

– they shelter some microorganisms (adsorbed on their surface or trapped inside) in the effluent for disinfection (barrier effect), which are then protected from the action of UV rays.

The minimum UV transmission for a media-filtered effluent is 55%. The minimum UV transmission for a membrane-filtered effluent is 75%. The minimum UV transmission for reverse osmosis effluent is 90%.

Secondary effluent filtration has a positive effect on UV performance. SS concentration should be less than 5 $mg \cdot L^{-1}$ and the solid diameter should be less than 10 μm to ensure readily an effluent disinfected at 10 fecal coliforms per 100 mL. In fact, considering the health hazard of parasites and knowing that a small dose is sufficient to cause an infection, it is essential to try to reduce as much as possible the load of parasites upstream of this last stage, which is disinfection, hence the importance of an optimal operation of secondary settling and the tertiary filter sizing.

Figure 8.32. *UV installation in a big wastewater reuse plant. For a color version of this figure, see www.iste.co.uk/gaid/reuse1.zip*

8.8.6.1.3. Hydraulics

The UV system should be designed for maximum flow at the end of lamp life. The UV system must also be optimized for medium and minimum flow rate to save on investment and operating costs. This can be done by dividing the flow into different channels and automating the system to control the UV dose. It is important that the flow rate in each channel be equal and measured to determine the appropriate UV dose to be applied. The flow must be laminar over the entire lamp network. Special attention must be paid to the sizes of the channel to avoid any short circuit of water around the lamp network.

8.8.6.1.4. Deposit formation on sleeves

UV lamps are protected from water contact by a quartz sleeve. The transparency of the sleeve is an important element for disinfection effectiveness, and several phenomena cause it to become opaque. These compounds absorb and reflect UV light and prevent it from entering wastewater. They are:

– viscous film due to grease;

– aluminum, iron, magnesium, calcium and phosphorus are the main compounds that cover the quartz sleeves protecting UV lamps;

– scaling (more intense with medium-pressure lamp technology, as the operating temperature is higher, resulting in an increased risk of mineral precipitation);

– algae growth;

– condensation inside the quartz sleeves.

The compounds likely to form deposits that impair radiation efficiency are mainly:

– calcium and magnesium in the case of scaling water; as the risk of scaling is further increased by the warming effect that must occur in the vicinity of the lamps;

– iron and manganese;

– SS and miscellaneous debris;

– residues of iron or aluminum salts used in coagulation.

A longitudinal flow of water along the lamps is likely to promote gradual precipitation of solids on the walls. Lamp deposits and fouling reduce the spread of UV radiation and involve periodic mechanical and chemical cleaning.

8.8.6.1.5. Emitted UV intensity

It depends on:

– the number of lamps in operation;

– the applied voltage (almost proportional intensity variation with the voltage);

– the state of fouling of the sleeves, hence the importance of regular cleaning;

– lamp degradation: the light intensity emitted by lamps decreases by 35% at the end of life. With internal irradiation, the quartz of the lamps suffers from a loss through solarization, a phenomenon which is one of the causes of aging.

This also occurs, but to a lesser extent, to sleeve equipment (quartz transmission could decrease by 50% after 20,000–30,000 h of service due to the effect of quartz solarization of the sleeves not including the effects of deposit formation).

8.8.6.1.6. Contacting

Influencing factors are as follows:

– contact time: it must be sufficient and compliant with supplier specifications;

– hydraulic flow conditions (mixing, preferential passages, etc.): they can be simulated with specific software.

8.8.6.1.7. Lamp aging

Lamps wear out over time due to the combined action of various factors:

– evaporation of the electron emitting electrodes (due to ignition without preheating);

– metal oxide deposit on the inner wall of the lamps (if no protective coating);

– surface attack of the quartz of the lamp inner wall by plasma, particularly reactive as highly charged in radicals or dissociated ionic species (the same if no coating);

– damage to the crystal lattice of quartz under the impact of UV photons (depends on the quartz quality).

Quartz sleeves also wear out, but to a lesser extent, because of:

– possible mechanical stresses due to the cleaning system and vibrations created by the passage of water (in reality, quartz is very resistant mechanically as long as it does not break under impact);

– damage to the crystal lattice of quartz under the impact of UV photons.

Lamp suppliers have developed internal lamp protection coatings that extend the life of lamps while limiting the harmful effects of lamp aging during operation. Currently, LP lamps have lifespans of easily 12,000–14,000 h.

8.8.6.1.8. Lamp lifespan

At the end of the guaranteed service life, the radiation emitted by the lamp in the germicidal wavelength range defined by the manufacturer should be equal to or greater than 80% of the measured radiation intensity for a new lamp. The lamp life is estimated from information given by the intensity sensors: a lamp is considered at the end of its life when, at maximum electrical power applied to the lamp, the intensity detected by the sensors is less than or equal to 80% of the intensity provided by a new lamp.

The lamp lifespan (low and medium pressure) varies between 12,000 and 14,000 h depending on the manufacturer. In recent years, the development of lamps less sensitive to operating conditions (switching on/off, cleaning, etc.) results in a lifespan of 14,000 h.

8.8.6.1.9. Quartz sleeves and ballasts

Quartz sleeves and ballasts are guaranteed *pro rata temporis* for a period of 10 years.

8.8.6.1.10. Lamp operation management

Switching on/off the lamps should be avoided as far as possible for the following reasons:

– When the lamp is switched on, there is a latency time (about 2 min) during which the lamp has not reached its steady state and therefore water is not disinfected.

– The number of switching on/off a lamp is limited (manufacturers generally recommend not to exceed four times per day, as each reduces the lamp life by about 10 h).

8.8.7. *Performance*

Like chemical disinfectants, UV irradiation does not kill microorganisms but inactivates them by removing their ability to multiply and infect the host. While how to determine this decrease in infectivity is known and easily achieved for viruses and bacteria, for parasites such as *Cryptosporidium* and *Giardia* it has been much more difficult to implement. In recent years, infectivity analysis methods on cell culture have emerged as alternatives to the various in vivo and in vitro techniques. Among the techniques to assess infectivity and disinfection efficiency for oocysts of

Cryptosporidium parvum, a cell culture method has been developed in recent years to get as close as possible to the environment in which the microorganism would be in a situation in vivo.

Cryptosporidium sensitivity between 276 and 287 nm is about 80% of that observed between 250 and 261 nm. Sensitivity of *Bacillus subtilis* viruses and spores is equivalent to 254 and 280 nm wavelengths. Finally, irradiation at 280 nm shows an efficiency very close to that observed at 254 nm for different coliforms and a virus. When disinfection is performed with LP lamps, it is easy to extrapolate the inactivation obtained in static at 254 nm with equivalent reduction dose obtained on a reactor. However, for medium-pressure lamps, it is necessary to relate the efficiency of each wavelength to the efficiency of the 254 nm wavelength. All these cases tend to show that irradiation at 254 nm with conventional LP mercury lamps is effective but can be matched by other wavelengths close to 280 nm.

Figure 8.33 shows that a dose below 30 mJ·cm^{-2} can inactivate up to 4 log of *Cryptosporidium parvum* parasites.

Table 8.26 provides some data on UV radiation dose–response for various microorganisms that are pathogens, indicators or microorganisms used in the validation of UV disinfection technologies.

In most cases, data are generated from LP lamps for which the lamp intensity can be measured and multiplied by the exposure time to obtain the applied dose.

Adenoviruses have a high UV resistance, and a UV dose of 165–186 mJ·cm^{-2} is required to inactivate 4 log with an LP lamp.

	Type of lamp	**UV dose (mJ·cm^{-2})**			
Microorganism	–	1 log	2 log	3 log	4 log
		Bacteria			
Coliforms	LP	<3	5	10	18
Escherichia coli	LP	<3	5	8	15
Aeromonas hydrophila	LP	1.1	2.6	3.9	5.9
Campylobacter jejuni	LP	1.6	3.4	4	4.6
Citrobacter diversus	LP	5	7	9	11.5
Escherichia coli	LP	6	6.5	7	8
Klebsiella pneumoniae	LP	12	15	17.5	29
Legionella pneumophila	LP	1.6	3.2	4.8	6.4

	Type of lamp	UV dose (mJ.cm^{-2})			
Bacteria					
Legionella pneumophila	MP	1.9	3.8	5.8	7.7
***Salmonella* spp.**	LP	<2	2	3.5	7
***Salmonella* Typhimurium**	–	2	3.5	5	9
Shigella dysenteriae	LP	0.5	1.2	2	3
Shigella sonnei	LP	3.2	4.9	6.5	8.2
Staphylococcus aureus	–	3.9	5.4	6.5	10.4
Streptococcus faecalis	–	6.6	8.8	9.9	11.2
Vibrio cholerae	LP	0.8	1.4	2.2	2.9
Yersinia enterocolitica	LP	1.7	2.8	3.7	4.6
Yersinia ruckeri	LP	1	2	3	5
***Bacillus subtilis* spore**	LP	24	35	47	79
Parasites					
Cryptosporidium parvum	LP	<3	5.8	12	22
Giardia lamblia	LP	<3	5.2	11	22
Viruses					
MS2 (phage) on *E. coli*	LP	15	32	60	85
Canine calicivirus	LP	7	15	22	30
Adenovirus type 15	LP	40	80	122	165
Adenovirus type 40	LP	55	105	155	186
Poliovirus 1	LP	7	17	28	37
Coxsackie B3 virus	LP	8	16	24.5	32.5
Rotavirus type 1	LP	–	–	16	36
Rotavirus SA-11	LP	7	13	20	35
Hepatitis A	–	5.5	9.8	15	25
Echovirus I	LP	8	16.5	25	33
Echovirus II	LP	7	14	20.5	28

Table 8.26. *UV doses (mJ·cm^{-2}) required for various microorganisms and log reduction*

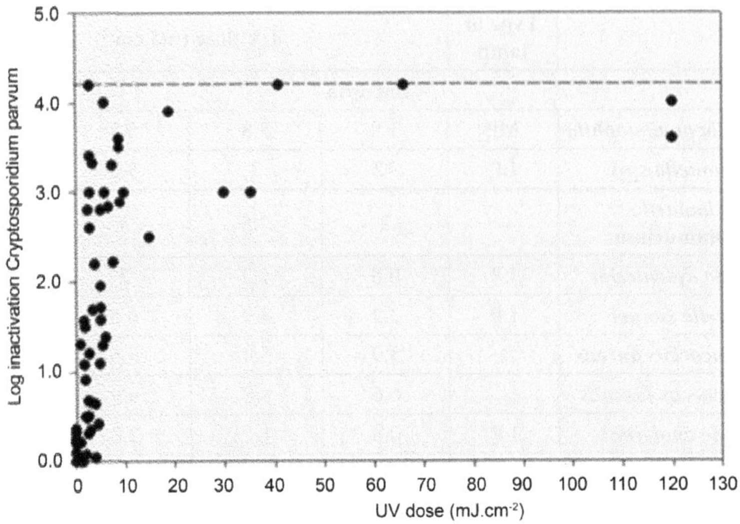

Figure 8.33. *Inactivation of* Cryptosporidium parvum

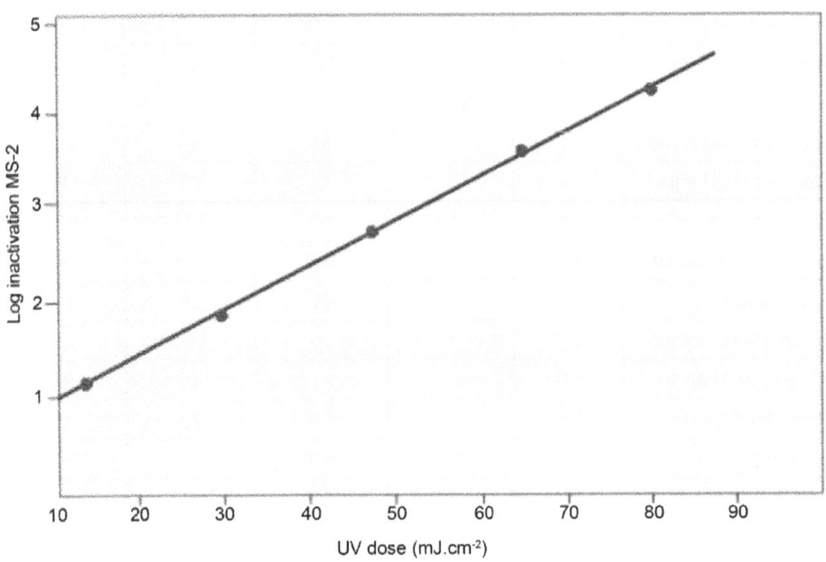

Figure 8.34. *Log inactivation of MS2 (on* E. coli*)*

Figures 8.34 and 8.35 show the inactivation doses for MS2 and various microorganisms in water.

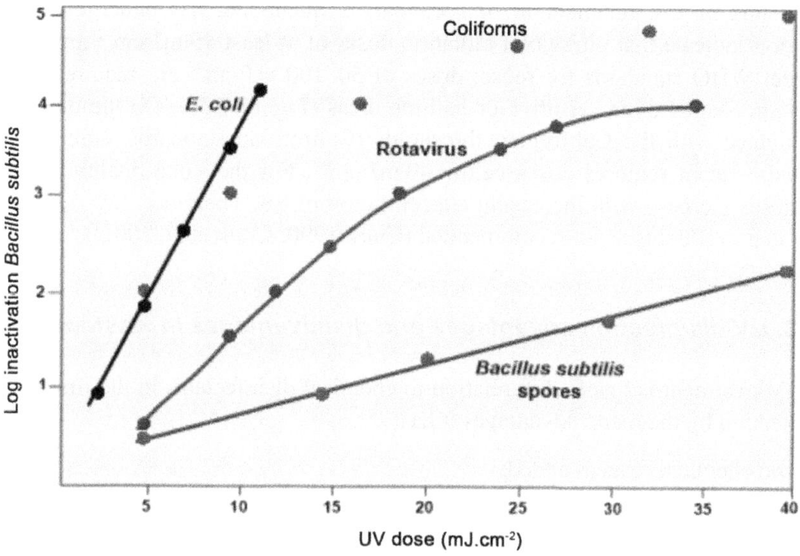

Figure 8.35. *Log inactivation of various microorganisms*

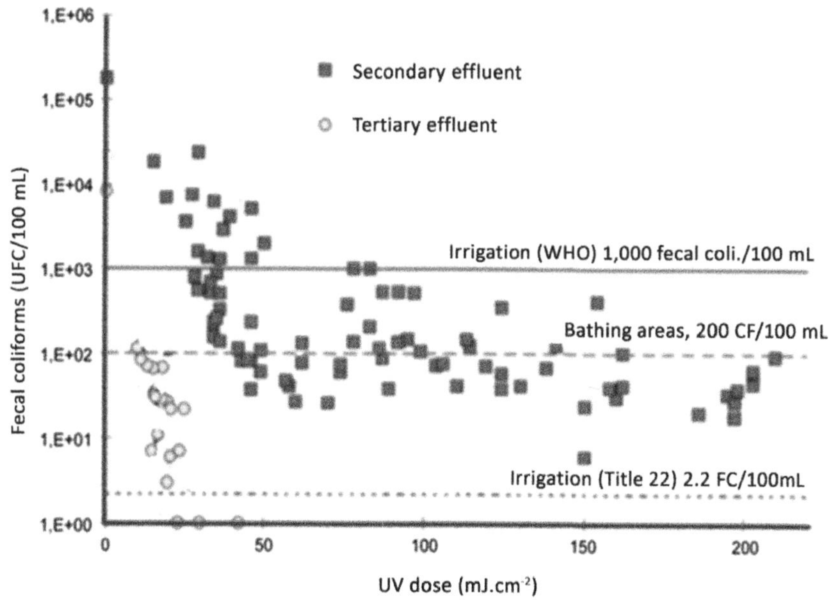

Figure 8.36. *UV disinfection performance to eliminate fecal coliforms for different quality standards*

Results of UV disinfection of secondary effluent that has undergone tertiary filtration indicate that ultraviolet radiation doses of at least 45 $mJ \cdot cm^{-2}$ are required to meet WHO standards for reuse; doses of 50–100 $mJ \cdot cm^{-2}$ are required to meet discharge standards in sensitive or bathing areas (Figure 8.36). On the other hand, compliance with the Californian thresholds (Californian standards Title 22) on a filtered effluent requires doses below 40 $mJ \cdot cm^{-2}$. For the same quality objective, UV doses increase with increasing concentration of SS. Therefore, UV disinfection of primary effluents is not recommended (Clark 1990; Craik et al. 2001).

8.8.8. *UV disinfection advantages and disadvantages in wastewater*

UV treatment is justified in relation to chemical disinfectants in the first stage of disinfection by the many advantages it has:

– no chemical reagent added;

– water does not have any taste or odor problems;

– no resistance to UV radiation is currently known in microorganisms;

– UV rays are not corrosive;

– very effective against parasites such as *Cryptosporidium* and *Giardia* and against various bacteria such as, in particular, legionella. Chemical disinfectants are only effective against parasites in high concentrations, which are difficult to consider for drinking water treatment due to the consequences linked to taste problems, secondary reactions, etc. UV technology (DRE 40 $mJ \cdot cm^{-2}$) becomes an alternative of choice and would fit into the concept of a multi-barrier treatment in which an optimal clarification and a possible refining stage are combined, on the one hand, and the addition of a second oxidant for network protection, on the other hand.

Cost: 0.04–0.06 €/m^3.

This solution has the following advantages:

– reduction of microorganisms and in particular parasites (*Cryptosporidium* and *Giardia*);

– no reaction by-products;

– no chemical reagents.

Secondary disinfection (chlorine or derivatives) will help protect distribution networks. In this case, the disinfectant demand should be lower than the demand observed during a conventional one-step disinfection.

Advantages	Disadvantages
– Significant reduction of the parasites *Cryptosporidium* and *Giardia*, much higher than the reduction obtained with chemical disinfectants, especially chlorine dioxide and chlorine if not preceded by ozonation: according to recent studies, it would be possible to inactivate 3–4 log (99.9–99.99%) of *Cryptosporidium* under typical implementation conditions. Regarding the efficiency on *Giardia*, 4 log elimination of infectivity for doses between 10 and 100 $J \cdot m^{-2}$. However, results often depend on the method of analysis for infectivity; – good bactericidal and virucidal efficiency; – physical process that changes the water chemical composition little and does not involve the formation of undesirable compounds encountered in chemical disinfection or problems with taste and odor; – no need for on-site storage of reagents and thus elimination of risks to staff and related safety problems; – pH-independent efficiency (as is the case for chlorine); – very short contact time (a few seconds) required directly in the irradiation chamber; – possibility of continuous UV emission control; – ease of implementation and operation.	– No residual power capable of preserving the good quality of the treated water during distribution and no possibility to directly assess the efficiency of treatment with residual measurement as in the case of a chemical oxidant; – maintenance-sensitive efficiency; – efficiency influenced by certain water quality parameters to be taken into account during project preparation and operation; – treatment more expensive than chlorine; the cost is directly proportional to the size of the installation; – fairly broad scope as a single disinfection treatment.

Table 8.27. *UV disinfection advantages and disadvantages*

Bromate regulations (10 $\mu g \cdot L^{-1}$) also provide an incentive for operators to reduce the ozone concentrations applied in interozonation and post-ozonation to limit the production of bromates. This decrease is reflected by a lower disinfecting effect of this step. They compensate for this decrease by increasing the final doses of chlorine or derivatives. The consequence would be a change in concentrations of by-products. The alternative solution that has developed in recent years is to integrate a first step with UV treatment which compensates for the ozone deficit without affecting the secondary chemical disinfectant demand.

Especially since UV radiation does not affect the various physico-chemical parameters (nitrates, nitrites, nitrosamines, ammonium, COD) or the physico-chemical characteristics of water at the operating dose of 40 mJ·cm^{-2}. Knowing that even a higher dose, up to 150 mJ·m^{-2}, did not cause nitrite formation.

8.9. Criteria for selecting a disinfection technique

The criteria for selecting wastewater disinfection processes and operating conditions are presented in Tables 8.28 and 8.29.

Criteria	Liquid bleach Chlorine gas	Ozone	UV	Organic peracids
Active disinfectant	$HClO^-$	O_3	λ = 254 nm	Peracetic acid Performic acid Perpropionic acid
Residual	Good	Nil	Nil	–
Cost	+	+++	++	+
Limitations	Most used, but often controversial process (potential formation of organochlorines, destruction of the flora present in lagoons or receiving environment, etc.)	Efficient process but requires effluent with good SS elimination. Formation of bromates	Efficient process but requires good pre-filtration and flow regulation. Well controlled, simple and effective	Process prepared in situ. Highly dependent on the supplier

Table 8.28. *Comparative criteria between the various disinfectants*

Chlorine has been used as a disinfectant for many years. Its effectiveness depends on water temperature, pH, type of injection, contact time, presence of interfering substances, chlorine concentration and form, as well as the nature and concentration of the target microorganisms. Bacteria are known to be less resistant to chlorine than viruses, which in turn are less resistant than oocysts and parasite cysts. The dosage required to disinfect at the desired level varies depending on the presence of constituents that interfere with chlorine efficiency. Typical constituents are organic matter, SS and ammonia. Organic matter in water may reduce chlorine availability for disinfection due to the oxidation of organic matter by chlorine. A higher dose of chlorine is required when water contains SS, which protect microorganisms from chlorine. Chlorine is a proven and effective chemical. However, long-term use and unregulated dosing can result in the generation of chlorinated organic compounds from the chlorination of residual organic matter in water. Such compounds may be carcinogenic. A dechlorination step is therefore important if water has to be discharged into larger water bodies.

Disinfection process	Objectives I < 10,000 TC/100 mL < 2,000 FC/100 mL	Contact time for objectives I	Objectives II < 100 FC/100 mL	Contact time for objectives II
Chlorine	3–8 mg.L^{-1}	30 minutes	8–10 mg.L^{-1}	30–45 minutes
Ozone	3–10 mg.L^{-1}	8 minutes	5–10 mg.L^{-1}	10 minutes
Chloramines	5–15 mg.L^{-1}	60 minutes	8–20 mg.L^{-1}	1–2 hours
UV	30–45 mWs/cm^2	50–100 mWs/cm^2	–	50–100 mWs/cm^2
PAA	5–8 mg.L^{-1}	17–40 minutes	10–20 mg.L^{-1}	20–60 minutes
PFA	2–5 mg.L^{-1}	10–15 minutes	5–15 mg.L^{-1}	15–30 minutes

Table 8.29. *Reduction comparison for different oxidants*

Ozone (O_3) is a powerful oxidant that can effectively oxidize organic substances. Ozone does not require any storage space, as it can be generated on site using the equipment needed to pass oxygen through high-voltage current. Its high oxidative power can destroy any microorganism in a few minutes. However, ozone must be generated on site and cannot be stored and transferred due to its unstable nature. The preparation of this product is a major energy consumer and requires more complex maintenance than a chlorination system. Ozone does not persist in water. In many cases, chlorination is performed after UV disinfection to maintain a residual chlorine and prevent microbial contamination during transport and storage of reused treated wastewater.

UV radiation is a physical disinfectant operating at a wavelength of 254 nm, which can penetrate the microorganism cell wall and be absorbed by cellular nucleic acids. This absorption stops cell replication and causes the death of microorganisms. UV disinfection does not produce any disinfection by-products and is less expensive than chlorine. It does require however the prior elimination of SS, which protect microorganisms. Turbidity and SS should be less than 5 NTU and 10 mg·L^{-1}, respectively. In many cases, chlorination is performed after UV disinfection to maintain residual chlorine and prevent microbial contamination during the transportation and storage of reused treated wastewater. Proper maintenance of the lamp and equipment is necessary for long-term use of UV light (Liberti et al. 2000; Mezzanotte et al. 2007).

Organic peracids such as PFA and PAA are two effective and reliable disinfection reagents in wastewater treatment applications (Table 8.30).

Escherichia coli reduction of 2–3 log is obtained at dosages of 1–1.2 mg·L^{-1} for contact times less than 20 min. For irrigation, the guide value of 10^3 per 100 mL can be respected under these conditions. With respect to quality impact, no by-product formation or ecotoxicological effects were observed. No effect on organic substances was observed, indicating a low potential propensity of PFAs to form by-products.

	Ozone	UV	PAA/PFA	Chlorine
Adenovirus	Good	Low	–	Good
MS2 coliphage	Good	Correct	Correct	Good
Enterovirus, poliovirus	Good	Good	Medium	Good
Cryptosporidium	Correct	Good	Low	Correct
Giardia lamblia	Correct	Good	Low	Correct

Table 8.30. *Compared efficiency of PAAs and PFAs with other oxidants for viruses and protozoa*

8.10. References

Tertiary filtration

Adin, A., Baumann, E.R., Cleasby, J.L. (1979). The application of filtration theory to pilot-plant design. *J. Amer. Water Works Assoc.*, 71(1), 17–27.

Amirtharajah, A. (1977). Optimum backwashing of sand filters. *J. Environ. Eng. Div. ASCE*, 104(EE5), 917–938.

Amirtharajah, A. and Cleasby, J.L. (1972). Predicting expansion of filters during back-wash. *J. Amer. Water Works Assoc.*, 64(1), 52–59.

Armon, R., Mandelbaum, R., Manor, Y., Campos, C., Gillerman, L., Salgot, M., Gerba, C., Klein, I. (2001). Secondary wastewater disposal for crop irrigation with minimal risks. *Water Sci. Technol.*, 43(10), 139–146.

Asano, T. and Cotruvo, J.A. (2004). Groundwater recharge with reclaimed municipal waste-water: Health and regulatory considerations. *Water Res.*, 38, 1941–1951.

Camp, T.R., Graber, S.D., Conklin, G.F. (1971). Backwashing of granular water filters. *J. Environ. Eng. Div. Proc. ASCE*, 97(EE6), 903–919.

Carman, P.C. (1937). Fluid flow through granular beds. *Chemical Engineering Research & Design*, 75, S32–S48.

Carman, P.C. (1947). Slip flow in granular beds. *Nature*, 160, 301–302.

Cho, A.G., Dodds, J., Santamarina, J.C. (2006). Particle shape effects on packing density, stiffness and strength – Natural and crushed sands. *Journal of Geotechnical and Geoenvironmental Engineering*, 132(5), 591–602.

Cleasby, J.L. (1993). Water science and technology. *Declining Rate Filtration, Water Science and Technology*, 277(8), 11–22.

Crittenden, J.C., Trussell, R.R., Hand, D.W., Howe, K.J., Tchobanoglous, G. (2012). *MWH's Water Treatment: Principles and Design*. John Wiley & Sons, New York.

Culp, R.L. (1977). Direct filtration. *J. Amer. Water Works Assoc.*, 69, 375–391.

Dawoud, M.A., Sallam, O.M., Mahmoud, A., Abdelfattah, M.A. (2010). Treated wastewater management and reuse in arid regions: Abu Dhabi case study water resources department. In *Environment Agency, 10th Gulf Water Conference*, Abu Dhabi.

Ergun, S. (1952a). Fluid flow through packed columns. *Chemical Engineering Progress*, 48(2), 78–89.

Ergun, S. (1952b). Determination of geometric surface area of crushed porous solids. *Analytical Chemistry*, 24, 369–388.

Ergun, S. and Orning, A.A. (1949). Fluid flow through randomly packed columns and fluidized beds. *Industrial and Engineering Chemistry*, 41(6), 1179–1184.

Exall, K. (2004). A review of water reuse and recycling, with reference to Canadian practice and potential. *Water Qual. Res. J. Canada*, 39(1), 13–28.

Exall, K., Marsalek, J., Schaefer, K. (2004). A review of water reuse and recycling, with reference to Canadian practice and potential 1. Incentives and implementation. *Water Qual. Res. J. Canada*, 39(1), 1–12.

Fasciolo, G., Meca, M., Gabriel, E., Morabito, J. (2002). Effects on crops of irrigation with treated municipal wastewaters. *Water Sci. Technol.*, 45(1), 133–138.

Fitzpatrick, C.S.B. (1990). Detachment of deposits by fluid shear during filter backwashing. *Water Supply*, 8, 177–183.

Gaid, K. (2002). Wastewater reuse for the industry. In *Conf. Environment – Industry*, Curitiba.

Gaid, K. and Bonne, P. (2003). Les technologies membranaires dans la réutilisation des eaux usées. *L'Eau, L'Industrie, Les Nuisances*, 125, 56–68.

Gaid, K. and Lanning, J. (2000). Wastewater reuse: The case of Durban. In *French-India Conf. on Environment*, Mumbai.

Gaid, K. and Le Poder, N. (1998). La réutilisation des eaux usées pour l'agriculture : critères de qualité et moyens techniques. In *IXe Congrès U.A.D.E.*, Casablanca.

Gaid, K. and Nugier, L. (2006). Traitement tertiaire des eaux usées en France. Le cas de l'usine de l'Hermitage (La Réunion). *L'Eau, L'Industries et Les Nuisances*, 322, 322–342.

Gaid, K. and Zeghal, S. (1998). Wastewater technologies reuse. In *Water & Environment Conf.*, Cairo.

Graham, N.J.D. (1988). Filter pore flocculation as a mechanism in rapid filtration. *Water Research*, 22(10), 1229–1238.

Hamoda, M.F., Al-Ghusain, I., Al-Mutairi, N.Z. (2004). Sand filtration of wastewater for tertiary treatment and water reuse. *Desalination*, 164(3), 203–211.

Hutchinson, W.R. (1976). High-rate direct filtration. *J. Amer. Water Works Assoc.*, 65(9), 292–298.

Kawamura, S. (1975). Design and operation of high-rate filters. *J. Amer. Water Works Assoc.*, 67(10), 535–549.

Koivunen, J., Siitonen, A., Heinonen-Tanski, H. (2003). Elimination of enteric bacteria in biological/chemical wastewater treatment and tertiary filtration units. *Water Res.*, 37, 690–698.

Kozeny, J. (1927). Uber Kapillare Leitung des Wasser in Boden. *Wien Akad. Wiss.*, 136, 271–306.

Lawler, D.F. and Nason, J.A. (2006). Granular media filtration: Old process, new thoughts. *Water Science and Technology*, 53(7), 1–7.

Leva, M. (1959). *Fluidization*. McGraw-Hill, New York.

Liao, C.W., Yu, J.H., Tang, Y.S. (2010). On-line full scan inspection of particle size and shape using digital image processing. *Particuology, Chinese Society of Particuology*, 8(3), 286–292.

McEwen, B. (1998). Indirect potable reuse of reclaimed water. In *Wastewater Reclamation and Reuse*, Asano, T. (ed.). CRC Press, New York.

Mitchell, C., Chong, J., Turner, A., Retamal, M., Carrard, N., Murta, J., Mukheibir, P., Moy, C. (2013). Darling Quarter Case Study. Building Industry Capability to Make Recycled Water Investment Decisions. Institute for Sustainable Futures, University of Technology, Sydney, for the Australian Water Recycling Centre of Excellence.

Mujeriego, R., Cazurra, T., Gullon, M. (2008). Wastewater reuse in El Prat de Llobregat, Barcelona. *Water and Sciences Technology*, 57(4), 567–574.

Ncube, P., Pidou, M., Stephenson, T., Jefferson, B., Jarvis, P. (2016). The effect of high hydraulic loading rate on the removal efficiency of a quadruple media filter for tertiary wastewater treatment. *Water Research*, 107, 102–112.

Ni, C.H., Chen, J.N., Tsai, Y.C., Chen, T.K., Chen, W.B., Chen, C.H. (2003). Recycling and reuse of wastewater from a new-developed community using sand filtration, ultrafiltration, and ozonation. *J. Environ. Sci. Health A: Tox. Hazard. Subst. Environ. Eng.*, 38(10), 2339–2348.

Ogoshi, M., Suzuki, Y., Asano, T. (2001). Water reuse in Japan. *Water Sci. Technol.*, 3(10), 17–23.

Okun, D. (1997). Distributing reclaimed water through dual systems. *J. Amer. Water Works Assoc.*, 89(11), 52–64.

Olivieri, A.W., Eisenberg, D.M., Cooper, R.C., Tchobanoglous, G., Gagliardo, P. (1996). Recycled water – A source of potable water: City of San Diego health effects study. *Water Sci. Technol.*, 33(10/11), 285–296.

Perez, S., Köck-Schulmeyer, M., Ginebreda, A. (2010). Wastewater reuse in the Mediterranean area of Catalonia, Spain. Case study of reuse tertiary effluent from a wastewater treatment plant at El Prat de Llobregat (Barcelona). In *Waste Water Treatment and Reuse in the Mediterranean Region*, Barceló, D. and Petrovic, M. (eds). Springer, Berlin.

Pollice, A., Lopez, A., Laera, G., Rubino, P., Lonigro, A. (2004). Tertiary filtered municipal wastewater as alternative water source in agriculture: A field investigation in Southern Italy. *Science of the Total Environment*, 24, 201–210.

Quinzaños, S., Strube, R., Dahl, C., Mujeriego, R. (2008). Helminth eggs removal by micro screening for water reclamation and reuse. *Water Science & Technology*, 57(5), 715–720.

Rodgers, S.E. and Lauer, W.C. (1992). Denver's demonstration of potable water reuse: Water quality and health effects testing. *Water Sci. Technol.*, 26(7/8), 1555–1564.

Sanz, J., Strube, R., Quinzaños, S., Dahl, C., Montoliu, I., Gracenea, M., Mujeriego, R. (2009). Helminth eggs removal in water reclamation on disk-filters. In *International Conf. on Disinfection of Wastewater*, Barcelona.

Sneed, E.D. and Folk, R.L. (1958). Pebbles in the Lower Colorado River, Texas: A study in particle morphogenesis. *The Journal of Geology*, 66(2), 114–150.

Stefanakis, A. (2015). Modern water reuse technologies. In *Handbook of Urban Water Reuse*, Eslamian, S. (ed.). CRC Press, New York.

Tchobanoglous, G., Burton, F.L., Stensel, H.D. (2002). *Wastewater Engineering Treatment and Reuse*. McGraw-Hill, New York.

Wen, C.Y. and Yu, Y.H. (1966). Mechanisms of fluidization. *Chemical Engineering Progress*, 62(4), 75–85.

Yao, K.M., Habibian, M.T., O'Melia, C.R. (1971). Water and wastewater filtration: Concepts and applications. *Environ. Sci. Technol.*, 5, 1105–1112.

Zhou, H. and Smith, D.W. (2002). Advanced technologies in water and wastewater treatment. *Journal of Environmental Engineering and Science*, 1, 247–264.

Disinfection

Antonelli, M., Rossi, S., Mezzanotte, V., Nurizzo, C. (2006). Secondary effluent disinfection: PAA long-term efficiency. *Environmental Science & Technology*, 40, 4771–4775.

Appels, L., Assche, A.V., Willems, K., Degrève, J., Impe, J.V., Dewil, R. (2011). Peracetic acid oxidation as an alternative pre-treatment for the anaerobic digestion of waste activated sludge. *Bioresource Technology*, 102, 4124–4130.

Azzellino, A., Antonelli, M., Canziani, R., Malpei, F., Marinetti, M., Nurizzo, C. (2011). Multivariate modelling of disinfection kinetics: A comparison among three different disinfectants. *Desalination and Water Treatment*, 29, 128–139.

Bach, R.D., Ayala, P.Y., Schlegel, H.B. (1996). A reassessment of the bond dissociation energies of peroxides. An ab initio study. *Journal of American Chemical Society*, 118, 12758–12765.

Baldry, M.G.C., Cavadore, A., French, M.S., Massa, G., Rodrigues, L.M., Schirch, P.F.T., Threadgold, T.L. (1995). Effluent disinfection in warm climates with peracetic acid. *Water Science & Technology*, 31, 161–164.

Bianchini, R., Calucci, L., Caretti, C., Lubello, C., Pinzino, C., Piscicelli, M. (2002a). An EPR study on wastewater disinfection by peracetic acid, hydrogen peroxide and UV irradiation. *Annali di Chimica*, 92, 783–793.

Bianchini, R., Calucci, L., Caretti, C., Lubello, C., Pinzino, C. (2002b). Intermediate free radicals in the oxidation of wastewater. *Research on Chemical Intermediates*, 28, 247–256.

Block, S.S. (2001). *Disinfection, Sterilization, and Preservation*. Lippincott Williams & Wilkins, Philadelphia.

Bolton, J.R. and Karl, G.L. (2003). Standardization of methods for fluence (UV Dose) determination in bench-scale UV experiments. *Journal of Environmental Engineering*, 29(3), 209–215.

Briancesco, R., Veschetti, E., Ottaviani, M., Bonadonna, L. (2005). Peracetic acid and sodium hypochlorite effectiveness in reducing resistant stages of microorganisms. *Central European Journal of Public Health*, 13, 159–162.

Bukhari, Z., Hargy, T.M., Bolton, J.R., Dussert, B., Clancy, J.L. (1999). Medium pressure UV light for oocyst inactivation. *J. Am. Water Works Assoc.*, 91, 86–94.

Cavallini, G.S., De Campos, S.X., De Souza, J.B., Vidal, C.M. (2013). Evaluation of the physico-chemical characteristics of wastewater after disinfection with peracetic acid. *Water, Air, & Soil Pollution*, 224, 1752–1763.

Chhetri, R.K., Thornberg, D., Berner, J., Gramstad, R., Öjstedt, U., Sharma, A.K., Andersen, H.R. (2014). Chemical disinfection of combined sewer overflow waters using performic acid or peracetic acids. *Science of the Total Environment*, 490, 1065–1072.

Chick, H. (1908). Investigations of laws of disinfection. *Journal of Hygiene*, 8(1), 92–158.

Clark, R.M. (1990). Modeling inactivation of Giardia lamblia. *Journal of Environmental Engineering, American Society of Civil Engineers*, 116(5), 837–853.

Collivignarelli, C., Bertanza, G., Pedrazzani, R. (2000). A comparison among different wastewater disinfection systems: Experimental results. *Environmental Technology*, 21(1), 1–16.

Craik, S.A., Weldon, D., Finch, G.R., Bolton, J.R., Belosevic, M. (2001). Inactivation of Cryptosporidium Parvum oocysts using medium- and low-pressure ultraviolet radiation. *Water Res.*, 35, 1387–1398.

Darby, J.L., Snider, K.E., Tchobanoglou, G. (1993). Ultraviolet disinfection for wastewater reclamation and reuse subject to restrictive standards. *Wat. Env. Res.*, 65, 169–180.

Dell'Erba, A., Falsanisi, D., Liberti, L., Notarnicola, M., Santoro, D. (2007). Disinfection byproducts formation during wastewater disinfection with peracetic acid. *Desalination*, 215, 177–186.

Dorn, J.M. (1974). A comparative study of disinfection on viruses and bacteria by monochloramine. Master's Thesis, University of Cincinnati, Cincinnati.

Falsanisi, D., Gehr, R., Santoro, D., Dell'Erba, A., Notarnicola, M., Liberti, L. (2006). Kinetics of PAA demand and its implications on disinfection of wastewater. *Water Quality Research Journal of Canada*, 41, 398–409.

Féliers, C. and Laplace, C. (2001). Évaluation de la dose dans un réacteur vertical en utilisant la biodosimétrie, l'actinométrie et le calcul CFD. In *2nd World Water Congress*, Berlin.

Finch, G.R., Haas, C.N., Oppenheimer, J.A., Gordon, G., Trussell, R.R. (2001). Design criteria for inactivation of cryptosporidium by ozone in drinking water, ozone: Science & engineering. *Journal of the International Ozone Association*, 23(4), 259–284.

Flores, M.J., Lescano, M.R., Brandi, R.J., Cassano, A.E., Labas, M.D. (2014). A novel approach to explain the inactivation mechanism of Escherichia coli employing a commercially available peracetic acid. *Water Science & Technology*, 69, 358–363.

Gaid, K., Girodet, P., Bord, G., Féliers, C. (2005). La désinfection aux ultraviolets : une alternative de choix. *L'Eau, l'Industrie et les Nuisances*, 282, 45–52.

Gehr, R. and Cochrane, D. (2002). Peracetic acid (PAA) as a disinfectant for municipal wastewater: Encouraging performance results from physicochemical as well as biological effluents. In *Proceedings of the US Water Environment Federation Disinfection Conference*. Water Environment Federation (Weftec), New Orleans, LA, 1, 182–198.

Gehr, R., Wagner, M., Veerasubramanian, P., Payment, P. (2003). Disinfection efficiency of peracetic acid, UV and ozone after enhanced primary treatment of municipal wastewater. *Water Research*, 37, 4573–4586.

Gehr, R., Chen, D., Moreau, M. (2009). Performic acid (PFA): Tests on an advanced primary effluent show promising disinfection performance. *Water Science & Technology*, 59, 89–96.

Gunther, L., Heinrich, R., Kurt, S. (1969). Process for stabilizing solutions of aliphatic percarboxylic acids. U.S. Patent, 3442937.

Gyürék, L.L. and Finch, G.R. (1998). Modeling water treatment chemical disinfection kinetics. *Journal of Environmental Engineering*, 124, 783–793.

Haas, C.N. and Engelbrecht, R.S. (1980). Chlorine dynamics during inactivation of coliforms, acid-fast bacteria and yeasts. *Water Res.*, 14, 1744–1755.

Heinonen-Tanski, H. and Miettinen, H. (2010). Performic acid as a potential disinfectant at low temperature. *Journal of Food Process Engineering*, 33(6), 1159–1172.

Hill, A.G. and Rice, R.G. (1982). Historical background, properties and applications. In *Handbook of Ozone Technology and Applications 1*, Rice, R.G. and Netzer, A. (eds). Ann Arbor, New York.

Howarth, J. (2003). Decay kinetics pf peroxyacetic acid (PAA) and hydrogen peroxide in a variety of water matrices, Modesto. *Enviro. Tech. Chemical services*, 3–12.

Jacangelo, J., Darby, J.L., Loge, F., Tchobanoglous, G., Heath, M., Swaim, P. (1995). Comparison of UV irradiation to chlorination: Guidance for achieving optimal UV performance. Report, Water Environment Research Foundation, Project 91-WWD-1.

Karpova, T., Pekonen, P., Gramstad, R., Öjstedt, U., Laborda, S., Heinonen-Tanski, H., Chávez, A., Jiménez, B. (2013). Performic acid for advanced wastewater disinfection. *Water Science & Technology*, 68, 2090–2096.

Kitis, M. (2004). Disinfection of wastewater with peracetic acid: A review. *Environ. International*, 30, 47–55.

Klenk, H., Götz, P.H., Siegmeier, R., Mayr, W. (2005). *Peroxy Compounds, Organic. Ullmann's Encyclopedia of Industrial Chemistry*. John Wiley & Sons, New York.

Koivunen, J. and Heinonen-Tanski, H. (2005a). Inactivation of enteric microorganisms with chemical disinfectants, UV irradiation and combined chemical/UV treatment. *Water Research*, 39, 1519–1526.

Koivunen, J. and Heinonen-Tanski, H. (2005b). Peracetic acid (PAA) disinfection of primary, secondary and tertiary treated municipal wastewater. *Water Research*, 39, 4445–4453.

Koivunen, J., Siitonen, A., Heinonen-Tanski, H. (2003). Elimination of enteric bacteria in biological–chemical wastewater treatment and tertiary filtration units. *Water Research*, 37, 690–698.

Kollu, K. and Ormeci, B. (2012). Effect of particles and bioflocculation on ultraviolet disinfection of *Escherichia coli*. *Wat. Res.*, 46, 750–760.

Lazarova, V., Janex, M.L., Fiksdal, L., Oberg, C., Barcina, I., Pommepuy, M. (1998). Advanced wastewater disinfection technologies: Short and long term efficiency. *Water Science Technology*, 38(12), 109–117.

Lefevre, F., Audic, J.M., Ferrand, F. (1992). Peracetic acid disinfection of secondary effluents discharged off coastal seawater. *Water Science & Technology*, 25, 155–164.

Li, H., Finch, G.R., Smith, D.W., Belosevic, M. (2001). Chlorine dioxide inactivation of cryptosporidium parvum in oxidant demand-free phosphate buffer. *Journal of Environmental Engineering*, 127(7), 594–603.

Liberti, L. and Notarnicola, M. (1999). Advanced treatment and disinfection for municipal wastewater reuse in agriculture. *Water Science & Technology*, 40(4/5), 235–246.

Liberti, L., Lopez, A., Notarnicola, M., Barnea, N., Pedahzur, R., Fattal, B. (2000). Comparison of advanced disinfecting methods for municipal wastewater reuse in agriculture. *Water Science & Technology*, 42, 215–220.

Luukkonen, T. (2016). New adsorption and oxidation bases approaches for water and wastewater treatment. Acta Universitatis Ouluensis. PhD Thesis, University of Oulu, Linnanmaa.

Luukkonen, T., Teeriniemi, J., Prokkola, H., Rämö, J., Lassi, U. (2014). Chemical aspects of peracetic acid based wastewater disinfection. *Water SA*, 40, 73–80.

Luukkonen, T., Heyninck, T., Lassi, U., Rämö, J. (2015). Comparison of organic peracids in wastewater treatment: Disinfection, oxidation and corrosion. *Water Research*, 85, 275–285.

Mattila, T. and Aksela, R. (2000). Method for the preparation of aqueous solutions containing performic acid as well as their use. U.S. Patent, 6492.

Mazzini, R., Pedrazzi, L., Lazarova, V. (2013). Production d'eau recyclée de haute qualité pour l'irrigation agricole à Milan. In *Jalons de la réutilisation de l'eau : les meilleures histoires de réussite*, Lazarova, V., Asano, T., Bahri, A., Anderson, J. (eds). IWA, London.

McNamara, P.J., Tucker, R. Burch, T.R., LaPara, T.M., Tan, D.T., Yan, M. (2011). Tertiary-treated municipal wastewater is a significant point source of antibiotic resistance genes. *Environmental Science & Technology*, 45(22), 48–62.

Merka, V., Sita, F., Zikes, V. (1965). Performic and perpropionic acids as disinfectants in comparison with peracetic acid. *Journal of Hygiene, Epidemiology, Microbiology, and Immunology*, 59, 220–226.

Mezzanotte, V., Antonelli, M., Citterio, S., Nurizzo, C. (2007). Wastewater disinfection alternatives: Chlorine, ozone, peracetic acid, and UV light. *Water Environment Research*, 79(12), 2373–2379.

Ragazzo, P., Chiucchini, N., Bottin, F. (2007). The use of hyproform disinfection system in wastewater treatment: Batch and full scale trials. In *Chemical Water and Wastewater Treatment IX*, Hahn, H.H., Hoffmann, E., Odegaard, H., (eds). LWA Publishing, London.

Ragazzo, P., Chiucchini, N., Piccolo, V., Ostoich, M. (2013). A new disinfection system for wastewater treatment: Performic acid full scale trials evaluations. *Water Sci. Technol.*, 67(11), 2476–2487.

Rajala-Mustonen, R.L., Toivola, P.S., Heinonen-Tanski, H. (1997). Effects of peracetic acid and UV irradiation on the inactivation of coliphages in wastewater. *Water Science & Technology*, 35, 237–241.

Rokhina, E.V., Makarova, K., Golovina, E.A., Van As, H., Virkutyte, J. (2010). Free radical reaction pathway, thermochemistry of peracetic acid homolysis, and its application for phenol degradation: Spectroscopic study and quantum chemistry calculations. *Environmental Science & Technology*, 44, 6815–6821.

Rossi, S., Antonelli, M., Mezzanotte, V., Nurizzo, C. (2007). Peracetic acid disinfection: A feasible alternative to wastewater chlorination. *Water Environment Research*, 79, 341–350.

Sanchez-Ruiz, C., Martinez-Royano, S., Tejero-Monzon, I. (1995). An evaluation of the efficiency and impact of raw wastewater disinfection with peracetic acid prior to ocean discharge. *Water Science & Technology*, 32, 159–166.

Santoro, D., Gehr, R., Bartrand, T.A., Liberti, L., Notarnicola, M., Dell'Erba, A., Falsanisi, D., Haas, C.N. (2007). Wastewater disinfection by peracetic acid: Assessment of models for tracking residual measurements and inactivation. *Water Environment Research*, 79, 775–787.

Shin, G.A., Linden, K.G., Sobsey, M.D. (2005). Low pressure ultraviolet inactivation of pathogenic enteric viruses and bacteriophages. *J. Environ. Engr. Sci.*, 4, 7–11.

Swern, D. (1949). Organic peracids. *Chemical Reviews*, 45, 1–68.

Veschetti, E., Cutilli, D., Bonadonna, L., Briancesco, R., Martini, C., Cecchini, G., Anastasi, P., Ottaviani, M. (2003). Pilot-plant comparative study of peracetic acid and sodium hypochlorite wastewater disinfection. *Water Research*, 37, 78–94.

Watson, H.E. (1908). A note on the variation of the rate of disinfection with change in the concentration of the disinfectant. *Journal of Hygiene*, 8(4), 536–542.

Yousefzadeh, S., Nabizadeh, R., Mesdaghinia, A., Nasseri, S., Hezarkhani, P., Beikzadeh, M., Valadi, A.M. (2014). Evaluation of disinfection efficiency of performic acid (PFA) catalyzed by sulfuric and ascorbic acids tested on *Escherichia coli* (ATCC, 8739). *Desalination and Water Treatment*, 52, 3280–3289.

Zuccato, E., Castiglioni, S., Mazzini, R. (2009). Pollution de l'environnement par les résidus de médicaments. *Ingengneria ambiental*, 7(8), 353–357.

Index

A, B

acid
  hypochlorous, 221
  peracetic, 271–274, 276, 278, 279, 304
  performic, 271, 272, 274–276, 278–280, 304
agriculture, 50
alkylphenol, 31
ammonia, 253
anthracite, 144, 159
aquaculture, 13
areas
  recreation, 13
  residential, 50
  swimming, 71
artificial lakes, 50
aryloxyacid, 30
Ascaris, 38
backwashing, 142
bacteria, 20
boron, 66
break-point, 226

C, D

carbamates, 29
cereals, 12
Chick's law, 216
chloramination, 253
chloramines, 212, 220, 221, 224, 226–228, 236, 249, 253–255, 257, 258, 260–263
chlorination, 134
  implementation, 230
chlorine, 207, 212, 213, 215, 216, 218–234, 236–250, 253–264, 268, 276, 282, 289, 291, 300–303
  active, 221–223
  application points, 225
  combined, 221
  demand, 226
  dioxide, 215, 218–220, 243–245, 247, 248, 254, 255, 258, 263, 264
  free, 221
  gaseous, 212
  total, 221
chloroacetamide, 30
cleaning
  street, 12, 50
  surface cleaning of roads, 51
climate change, 1
Clostridium
  perfringens, 36
  sulfite reducing, 36

coagulant, 147
coliforms
  fecal, 39
  total, 39
compounds
  perfluorinated, 32
  polybrominated, 32
contact tanks, 223, 228, 229, 232
crops
  food crops consumed raw, 96
  industrial, 12
  non-food, 97
Cryptosporidium, 38, 142, 196
Ct notion, 215
cycle times, 185
demand
  biological oxygen, 20
  chemical oxygen, 20
design of tertiary filtration, 193
diagram of floc diffusion, 183
dichloramine, 227
directives européennes, 86
disinfection, 29, 207
  by-products, 136
  tanks, 226
  UV, 281
  with
    ozone, 263
    performic acid, 275
drought, 3
drug
  molecules, 29
  residues, 30

E, F

economic comparison, 128
effect
  germicidal, 212
  residual, 212
effectiveness of various disinfectants, 213
electrical conductivity, 63
electrochloration, 234
emitted UV intensity, 293
endocrine disruptors, 29
Enterococcus, 35
enterovirus, 37
environment
  agricultural, 12
  industrial, 12
  urban, 12
Escherichia coli, 34
expanded
  clay, 150
  shale, 150
expansion of the material, 168
fecal
  bacteriophages, 210
  streptococci, 35
fields/courses
  golf, 50
  sports, 12
filter(s)
  backwashing conditions, 168
  dual media, 141
  high-rate, 142
  mobile, 179
    material, 197
  rotary drum, 198
  single-layer, 141, 176, 180
filtration, 134, 141
  fast gravity, 141
  gravity
    with reagents, 147
    without reagents, 142
  mechanical, 141
  on granular media, 141
  single-media, 152
  tertiary, 141
fire
  circuits, 12
  hydrants, 51
fishing grounds, 50

flocculation, 188, 189
floor
  Leopold type, 188
  nozzle draining, 188
fodder, 12
forests, 12
fountains, 12
fruit trees, 12

G, H

Giardia, 142, 196
  lamblia, 38
gravel, 151
green
  belts, 50
  spaces, 12
hardness, 151, 152, 156
health protection, 135
heavy metals, 21, 66
helminth eggs, 38
hormones, 30
HR filters, 182
hypochlorite, 221
  sodium, 212

I, L

illegal drugs, 29
injection mode and point, 218
interstices, 160
iron, 148
irrigation
  flood, 56
  furrow, 56
  landscape, 9, 50
  localized/local, 57, 107
  restricted, 106
  root, 57
  sprinkler, 57
  unrestricted, 103
lamps
  "medium and high-pressure", 287
  low-pressure, 286, 294, 296
Leopold system, 180
Leptospira, 34
liquid bleach, 220

M, N

market gardening, 12
media height, 142
Mediterranean countries, 115
Middle East, 115
mixing
  concrete, 51
  in tanks, 51
monochloramine, 220, 227, 228, 253–258, 260
nitrogen
  ammoniacal, 26
  global, 20
nozzles, 144, 152, 162, 173, 178, 180, 185, 190
nurseries, 50

O, P

organic
  carbon
    dissolved, 22
    total, 22
  peracids, 271
organophosphorus, 29
ozone, 207, 212, 215, 216, 218, 219, 255, 258, 263–270, 276, 281, 301, 303
parabens, 32
parasites, 20, 211
particle retention, 149
pastures, 12
pathogens, 19

performances
  of chlorine dioxide, 247
  of disinfection
    chlorine, 236
    with ozone, 268
  of gravity filters, 196
  of monochloramine, 256
pesticides, 29
phages of *B. fragilis*, 43
phenylurea, 29
planning, 127
play areas, 50
porosity, 148
power plants, 50
pozzolan, 150
preparation of chlorine dioxide, 244
pressure drop, 142
priority substances, 29
production of drinking water, 12
program(s)
  awareness, 129
  monitoring, 136
  public information, 129
protozoa, 33
  enteric, 38
public gardens, 50
pumice, 144

Q, R

qualities of a good disinfectant, 219
radiation
  UV, 134, 207
rate/velocity
  filtration, 142
recharge
  direct, 16
  groundwater, 12
regulations, 85
  French, 87
renovation of areas, 50
residential use, 88
residual chlorine, 229
restoration of drinking water, 13
retention capacity, 142, 144, 165, 167, 180, 183, 185
reuse
  direct, 16
    drinking water, 8, 15
  indirect, 16
  projects, 127
  unplanned, 9, 16
rotavirus, 38

S, T

salinity, 61
Salmonella, 34
sand, 143–146, 150, 152–154, 157–159, 163, 164, 166, 167, 174–176, 180, 182, 189
size
  effective, 144
  of the media, 142
sludge, 173, 174
snow making, 50
sodium adsorption rate, 64
spectrum of waves
  electromagnetic, 281
sprinkling, 10
Standard for Environment Reuse, 112
sulfonylurea, 30
supply of water features, 12
suspended solids, 20
Tænia saginata, 39
toilet flushing, 12
total phosphorus, 20
treatment
  conventional, 136
  plants, 137
  processes, 136
  quaternary, 134
  tertiary, 133
triazines, 29

trichloramine, 227
Trichuris, 38
turbidity, 141, 142, 145, 147, 158, 161, 168, 172, 175, 176, 189–191, 194, 196, 218, 292
type of media, 142

U, V

U.S. Environmental Protection Agency, 108
under pressure, 141
uniformity coefficient, 153
use(s)
  drinking, 52
  non-drinking, 52
  restricted, 58
  unrestricted, 58
UV dose, 284, 285, 290–293, 296
velocity
  minimum fluidization, 170
Vibrio cholerae, 34
virus(es), 20
  enteric, 37

W

washing
  floor, 50
  of vehicles, 12
wastewater
  quality classes, 97
  recyling, 16
  transmittance, 291
water
  bathing, 10
  cooling, 12, 50
  process, 12, 50
  sports, 71
  stress, 2
  supply
    drinking, 9
    ground, 9
  Unit Environmental Health Directorate, 113
watering
  of private gardens, 12
  of public parks, 12
wetlands, 13
World Health Organization, 86

Printed and bound by CPI Group (UK) Ltd, Croydon, CR0 4YY
13/09/2026
14961049-0006